IAN O'ROURKE

MEMORIES OF A TRACTOR MAN

Extra copies of this book are available
from the author by contacting via e-mail
at: ian.gwennie@gmail.com

IAN O'ROURKE

MEMORIES OF A TRACTOR MAN

First published in Australia in 2022 by Probert Consulting

Cataloguing in Publication Data is available from the Australian National Library.

ISBN: (trade paper-back) 978-0-6454178-4-5

ISBN: (E-publishing version) 978-0-6454178-3-8

Website: www.wurugi.blogspot.com.au
Trade enquiries: probertconsulting@bigpond.com

Cover Design: Studio Ink, Bendigo Victoria
Enquiries: hello@studioink.com.au

Cover Photo: Brandon Brooks, South Australia

Co-author: Terry L Probert

Printing: INGRAM

IAN O'ROURKE

MEMORIES OF A TRACTOR MAN

To Gwen my wonderful wife, this story is for you.

For over thirty-five years you have encouraged and lifted me whenever the going got tough. Even after a number of old friends suggested I write a book about my life, which at the time, I thought an impossible task, you were with me at the coalface, cajoling and encouraging me to put my memories into print. And, as you are so often right, this book is proof anything is possible.

Thank you, and now it's done, once again it is time to savour our quiet life on the Gold Coast.

FOREWORD

My memories of Ian O'Rourke are integral to those exciting and formative years in Ford Tractor Operations in Australia in the 1970s and 1980s. In that era, the ambience was predominantly one of a cohesive family, working to achieve common goals through our staff and a tight knit dealer group. The focus of all activity was, obviously, the achievement of sales objectives of both Ford tractors and "The Blue Line" vendor-sourced farm machinery. Concurrently, there was a clear undercurrent and philosophy of encouraging the highest standards of personal development from staff and dealer personnel, parallel with the achievement of sales objectives.

It can be characterised as a cohesive family environment, where personal idiosyncrasies, strengths and weaknesses were harnessed for the common good. Evidence of this diversity was the wide variety of backgrounds and experience of Ford T&E staff. We were made up of a hotchpot of identities, which would never be acceptable within the academic driven, narrow focussed business environment of the modern era. From memory we had these "square pegs" that fitted into the Ford T&E family: shearer, PhD Zoology, policeman, butcher, motor mechanic, farmer, various academics, earthmoving plant operator, and most every other life skill in between.

There was also a strong business sentiment of exclusivity and independence. We were not responsible to Ford Australia, other than the general sharing of resources such as the disciplines of Finance, Manufacturing, Legal, Parts support and Industrial Relations. But we always retained our independence in choosing to deviate from these corporate "restraints" when the unique requirements of our business model dictated. Conversely, we were more closely linked to the Ford Tractor world, reporting directly to World Headquarters in USA, with regular contact with personnel and management from there and throughout the world. This sharing of broad business philosophies and personal contacts on an international scale instilled a unique perspective to our role in Australia.

And in the middle of this unique corporate activity, Ian seemed to wear quite a few different hats. I first knew him when he was Ford T&E

Service Manager, battling the misty world of differing agendas: buyers, dealers, Ford Zone staff, product aberrations, Ford US and UK dithering, all told in the following pages. As Northern Region Service Representative one of my first jobs was to visit a Darling Downs' farmer who insisted Ford should buy him a new shed. It turns out he had left his old Ford 8000 tractor in gear (no Dual Power) and mice had devoured the wiring causing it to start and demolish the somewhat flimsy shed. Ian's response was a curt "No Way"!

And while balancing the often conflicting priorities of his Service Manager responsibilities, Ian was a prominent member of T&E's informal operating committee. He had essential involvement in all major product and policy decisions. In this capacity he was also our General Manager Noel Howard's "Mr Fix It" and was integral to the immediate and longer term viability of the operation. It is not an exaggeration to state that he was Noel's "right hand man" and a constant force in the 70s and 80s at Ford T&E.

This was an exciting and dynamic period that saw many innovations in the industry, initiated by Noel Howard's progressive and inspirational leadership. Most notable were the industry wide mandating of Roll Over Protective Structures

(ROPS); the world wide introduction of plastic fuel tanks; the uniquely and challenging local Ford 8401 tractor program; support of the local agricultural implement manufacturing industry through "The Blue Line" range; and numerous other projects Ian covers in this book.

While these initiatives were integral to the business, this era also included many diverse activities. These included new model introductions at Queensland's Gold Coast, TV endorsements of Ford TW Series by Sir Jack Brabham and even celebrity race sponsorship at the Australian Formula 1 Grand Prix. Exciting, stimulating and busy times!

Regrettably, with the disposal of Ford Tractor's heritage, innovation, reputation and significant market presence into the New Holland debacle, Ian became one of many loyal and capable personnel who were thrown onto the international corporate scrapheap. But he survived the chaos and never lost his empathy with his former work colleagues, nor compromised his integrity by lowering his standards of loyalty, honesty and basic common sense.

In retrospect, they were unrepeatable times, filled with exciting adventures and great people; never to be repeated in the impersonal corporate cesspit of these later times. As Mary Hopkins sang so poignantly: "Those were the days, my friend. We thought they'd never end". Regrettably they did, but they can at least be relived in Ian's wonderful stories in this book.

David Hosking

December, 2021

THE BOY FROM BURRABOI

About the middle of 1939, my world didn't extend much further than our small weatherboard house. It stood only fifty metres away from the tiny, unmanned Burraboi Railway Station in New South Wales. Not quite a railway brat, but I wasn't a farm kid either, not yet anyway. Listening to the radio and overhearing my parent's conversations, I began to realise there was a great big world out beyond the fences of the town. I became excited, dreaming about a bigger life and what it may hold for me.

After reading stories about the Flying Scotsman and other trains in one of Dad's books, I'd dream about living alongside one of those interstate lines. On my bed at night, the movies playing in my mind would have our little house rattling and rocking as a freight train snorting soot and steam strained its way past. It wasn't to be for a kid like me, though. All I could do was dream about laying there at night listening to the rails singing their song of approach. Instead, the only trains that worked our line, were steam trains with the power of a pushbike.

Even though they were slow, hissing and chugging their way between towns, these small engines performed a serious function for farmers along the line. On a Saturday, when his fat-lambs were ready for market, Dad would walk them to the Burraboi railway stock yards. There he would draft and pen them, so they were ready for loading when the train arrived at 3.00 pm, on Sunday. However, for the lambs it was a long journey until they reached Melbourne's Newmarket Saleyards. Their condition understandably poor from the journey, they often sold too cheaply and our income suffered.

Burraboi was one of many watering and coal stations along our line, a place where trains would load sheep, pigs, cattle, and wool. These were places where the crew could replenish it with coal and water before making it to the next stop, where the process would often be repeated. Situated on the rail line between Echuca in Victoria, and Balranald in New South Wales, its early importance had long since faded with the introduction of bigger and faster trains.

Ours was a spur line and ran for a distance of about two hundred kilometres. Other station towns along the line have been affected in similar ways and yet remain memorable for the unique character of their names. For a city dweller, towns like Bunnaloo, Womboota, Caldwell, Yallakool, Wakool, Burraboi, Jimaringle, Niemur, Dhuragoon, Moulamein, and Balranald must have seemed strange. Hard to pronounce correctly and little more than a rushed stop at a siding to some, they were the beginning of life's journey for people like me. Over time and with better roads the progress of road transport increased the pressure on these little communities causing people to move away, shops to close. Eventually these towns started looking unloved and rundown. It wasn't long before the line between the last two stations was closed in 1986.

From memory, I can only remember Wakool, Moulamein, and Balranald as having railway staff based on site. However, crews of gangers that serviced and maintained the tracks, lived in railway houses alongside the sidings. Often these were little more than shanties, and life for the wives of these men would have been hard. Even more difficult than for my own mother, whose time in Burraboi could not have been easy.

Our trains huffed and chugged their way along the rails every week. Often transporting sheep and other primary produce to market in Melbourne, or bringing basic supplies to farming families and other people living within a reasonable distance of the station. Although each train had a carriage for carrying passengers, in those days transport of any kind was difficult to afford. I can remember feeling sorry for travellers who after boarding the train in Echuca, would have to sit for hours tolerating a painfully slow journey. Waiting in a non-air-conditioned carriage as the train stopped at most stations required patience and resolve. Because the crew would often end up shunting various rolling stock off line and hook onto other cars and wagons, if you lived in Echuca and wanted to see a specialist doctor or visit family, all of this shunting made a journey to Bendigo last a considerable time.

Mum and I made the yawningly long journey twice and, it took over eight hours to go from Echuca to Wakool. One thing that sticks in my memory happened during the first journey. As boys who have been told not to, I had my head out the carriage window watching the scenery flash past and a hot cinder from the steam engine lodged in my right eye. It is an easy memory to keep, because when I look in a mirror and if I turn my eyes hard to the left, I can still see the scar on the pupil. It sure was painful and as I should have known better, there was some care, but little sympathy from Mum. A great lesson was learned on that trip.

I'm not sure how old I was, but I do know that seeing the trains started me wondering about the bigger world the adults talked about. It didn't take long before I began dreaming about those towns the trains passed through and the cities at the end of the line. My young mind raced with possibilities, planning how someday I might find a way to ride one of those passing trains and find out.

Back then, Dad was managing a farm of 1,200 acres, located four miles East of Burraboi. Every morning his routine began with a daily journey by horse and gig to the property. There the principal farming business was growing sheep for wool, and producing fat-lambs for the Melbourne markets. The farm had access to irrigation channels which had been constructed by the New South Wales Government. However, it like many other properties in the area only had the horses and their equipment to cultivate the land. At my young age, farming in this way was normal. Nary a tractor in sight.

At least in summer the channel was great, for it offered us a release from that indomitable farming routine. We could laze on the banks and swim in its cool water. We always had an opportunity to catch redfin, or bring tadpoles home and watch them as they grew into frogs. We didn't have much, but what we had we valued, even if it was just listening to chugging steam trains labouring past.

Around the same time each year, I would walk out into the field and find Dad ploughing with our three draught horses, preparing land to seed with grasses for grazing. Scotty, Bonnie & Jean were their names. Scotty was a cunning old horse; he had a habit of holding back when working, leaving Bonnie and Jean to take the load. For Dad it was frustrating and although he didn't curse often, even on the best of days, Scotty caused dad to bring forth profanity.

After a few months of working on the farm, Dad was able to purchase materials to build a house on the property. However, as Australia was still recovering from the Great Depression, supplies at that time were hard to obtain, and making things more difficult, the country was bracing for World War II to begin. Therefore, construction of our home was painfully slow.

Building the walls posed a problem, but Dad was able to enlist the help of a chap and with a couple of other mates, it wasn't long before they found a way to get the job started. Using the Murray pine floorboards that Dad had bought, they built framing to make walls. The boards stood upright and approximately two feet, or sixty centimetres apart. Fully braced on each side and with poles anchored at forty-five degrees, the framing stood solid.

Now it was time to mix up a "pizzey-mud" potion using clay, water and a quantity of straw, which they rammed down between the upright floor boards before leaving it to dry. Within a day or so the floorboard framework would be moved to a new section and the process repeated. Often framing would be inserted, providing openings for the later fitting of doors and windows. Like many others during this period, one of the biggest problems facing the home builder was obtaining materials. I remember the lengths Dad had to go to, to source enough corrugated iron for the roof. That alone took considerable time and held up the project.

To get water to the building site and to service other needs on the farm, Dad would harness one of our draft horses to a Furphy water cart. The Furphy became indispensable to us at this time, as we and most of New South Wales were struggling through the middle of a two-year drought. He would drive the cart down to the channel running through the farm, there he would bucket up water from it until he filled the tank, taking it to where it was needed most.

For Mum, and because of the drought and our tanks being dry, the most important place for the water cart was close to the house. When not employed somewhere else on the farm, that's where it stood. This way we could easily fill our two ten litre buckets that we used to carry the water inside. My parents insisted that all of the water coming into the house from the cart had to be boiled for at least ten minutes, sterilising it. Mum explained that the process of boiling water would kill all of the bacteria and other pathogens, making it safe to use. I remember her telling me that only a few generations before, people on the goldfields and in other overcrowded areas, had been dying of cholera and other water borne disease. All of which may have been avoided if they had understood the benefits of sterilisation. While we boiled it for drinking, hot water required for our washing and bathing was heated in a bucket that sat on the edge of the stove.

Our Furphy water cart had always intrigued me and because it had become so indispensable, in later life I often came across one of them tucked away in the back of a low roofed shed, or abandoned to weather away on the farm scrap heap. There, I would take time to stare and just appreciate them. Seeing that galvanised iron cylinder capped with cast iron ends, always took me straight back to the farm of my childhood. While the cart's cast iron wheels were different to the wooden ones on every other machine on the farm, it was the words cast into the rear end cap that captured my imagination. Even now when I happen on one, I love to trace my fingers over those words just as I did as a child.

The inscription read, "Good, Better, Best, Never Let it Rest, Till Your Good is Better and Your Better Best."

That simple message from the back of a tank by John Furphy and Sons of Shepparton Victoria, has always stood me in good stead. Not only for my career, but throughout life. It left such an impression on me, that I wish that many more of today's businesses and their workers held the same view.

By 1940, World War 2 was in full swing, and was something I could not fully comprehend. Life with my parents was still good and with plenty of home-grown food we never went hungry. However, at night listening to the old battery radio, the news coming from Melbourne's 3DB and 3LK radio stations had my parents worried. I would lay on my bed listening as they talked about the last war and cogitated on the effects of this one.

Although we only lived six miles from the small township of Wakool, there was no possibility for me to attend school there. Instead, arrangements were made for correspondence lessons to be mailed to the Post Office in Wakool. These lessons came from Blackfriars Correspondence School in Sydney. Each Saturday, Dad would put our pony Frisky into the gig and we would drive into Wakool to pick up my lessons. I had five days to complete them, then it would be back to Wakool. We'd pick up the new lessons and return post the completed set. Within three weeks the first set of lessons would return with numerous corrections and suggestions showing where I had made errors.

The trip into town was a good outing, we would pick up groceries and sometimes a small treat or two from Les Tynan's General Store. I loved those days with Dad, we had time for talk and allow any troubles of the week to melt away while we explored our minds.

Being war time, petrol was in very short supply, and like many other families in our district, we did not own a car until 1945 when dad purchased an old but sound 'A' model Ford.

Because the country had to put everything toward the war effort, rationing of essential supplies like petrol would continue until the situation normalized. Government coupons were dependent on a very strict criterion and were diligently distributed to every Australian household. These coupon books allowed us only four gallons, or just under eighteen litres of fuel per month, and for us, it meant the Ford could only be used for special occasions. It's hard to imagine having those limits on our fuel supplies today.

I can clearly remember one Saturday. We all got into the Ford, drove through Burraboi, then on toward a property owned by Alf Adams, his wife and three children. The reason for our visit was to attend the eldest boy's birthday. It's funny what you remember, but while we were playing in the yard, Mrs Adams came out with a bowl full of red cubes and all of the kids started grabbing one.

Encouraged by the others, I took one from the bowl too. Right away, my hand felt so cold it reminded me of the time I touched frost that had settled on the rails of the railway line. However, watching and learning the others, I found the ice-block was something quite soothing and refreshing to suck.

The Adams kids started laughing and telling me the cubes were ice blocks. However, for the life of me, I couldn't understand how they were made. We never had a machine that could do anything like that at home.

Later we went into the house and I was shown this large white machine. Mrs Adams took a lot of pleasure telling me it was an Electrolux refrigerator and it used kerosene and a flame to make it work. I understood kerosene and like everyone who lived away from major transmission lines, we were all using it in lamps for lighting our homes. Settling onto remote farmlands had limitations city folk could never understand.

However, turning heat into cold had me curious and I wanted to know more. For me, understanding latent heat and getting 240 Volt electric power to our farm, would be some time away. During war time, many things were very difficult to obtain and it would be two years after our day at the Adam's house before Mum and Dad were able to buy a 'fridge.

Until that time, mum would keep perishable food in a "Coolgardie Safe." Although not as effective as a fridge, at least it stopped flies settling on our food. A Coolgardie looked like a small box built on a hollow tin frame, it stood about three-foot-high and two-feet-square. On the top was a tin container, four inches, or about 100mm deep, which was for holding cool water. Stretched hessian cloth filled the sides and a side hinged door. More strips of hessian were weighted to hold them down into the water at the top, then be draped over the sides. Capillary action would cause the water in the container to slowly travel up the hessian strip until gravity would draw it over the edge and down the into the hessian on the door and sides. Most of the water in the hessian would evaporate in the warm air and, using the same principal as the fridge's latent heat, it resulted in a cooling of the contents. Although not as efficient as a refrigerator, it was the only way we could keep our perishable food for a bit longer.

For the time we were lucky to have an operating telephone line. Supported by natural timber poles, exposed copper wire carried our signal to the switchboard, in Wakool. There the operator would ask who we wanted to talk to and then by moving flexible wires with connectors at each end, they would link us to the person we wished to converse with. On our party-line, which was shared by four other farmers it would be possible to unwittingly break into someone's conversation. So, it was also possible for the four other people to listen in to your conversations too, maybe the people running the phone company thought privacy for country people was less important, or maybe they considered us too polite to be nosey.

We were running over a thousand breeding ewes on our farm and, to fit in with lambing, shearing would commence around July each year. Gus Petski from Barham was our contractor and he would bring his two-stand petrol driven shearing plant as our shed didn't have a big capacity. After mounting his rig into place, he would hire another shearer to help out.

One of our neighbours was a wool classer, and as soon as the sheep were relieved of their fleece, he would begin dividing various grades of wool into separate bins, ready for pressing into bales. The shed bustled with sweaty action as our two shearers fought for supremacy. The clatter of hooves on wooden boards challenging the noise of amazing machinery was matched only by the banter of the team of people getting the work done.

In a way a shearing shed is a bit like the Olympics and everyone is a competitor. However, it's the wool from each ewe that fought hardest for the prize, because only the price we received at a wool sale in Melbourne would prove the value of our work for the year.

The shearers stayed with us in our home during shearing and at night Gus would produce a pack of cards. They were something I had never seen before and over their time with us Gus showed us how to play various card games. He also taught me three card tricks that I'm still able to remember today and like Gus, I love teaching them to other young people too.

With the war still dragging on, it was very difficult to be able to buy any type of games including something small like a pack of cards. When Gus left, I endeavoured to make a set of playing cards from scrap cardboard utilising scissors and my coloured pencils. Needless to say, my full pack of fifty-two was almost a metre wide and impossible to play with.

Growing older, I yearned to own a pushbike and would often be found pestering my parents for one. Finding an old frame and wheels hanging in the shed, Dad oiled the chain and pedals, adjusting the seat and polishing the frame until he'd patched it up. Now that it was done, it was time for me to learn to ride. With my feet on the pedals, a death-grip on the handlebars and Mum watching from the kitchen window, holding the back of the seat, Dad pushed me around and around the front yard. After much fatherly patience and more circuits later, it finally happened. I found my balance and what a feeling of freedom it was to be riding on my own. Dad was laughing and puffing, and before getting his breath back, he told me that having mastered two wheels, it was now time for me to learn to ride a horse.

Some days later, he put a saddle on old Scotty, lifted me up, said hang on tight. He must have got quite a laugh when he cracked the stockwhip behind the horse. Hold on was right, Scotty and me were off, racing over a rough paddock for about four hundred metres of sheer childhood and equine terror.

My trainer, hands on hips and half grinning, half laughing and sucking in great gulps of air at our exploits, grabbed Scotty's bridle as we made our way back, "there you go Son," he said, now you can ride a horse."

He helped me down and whistling to himself, led Scotty back to his paddock.

One day I found a new hobby. In some soft ground near our mud house. I noticed a couple of holes in the ground, which were about half an inch in diameter and covered with a silvery looking lining. Being an inquisitive kid, I dug up this large and curious looking spider. It was dark in colour and I was moving in quite close, until it stood up on its back legs. A show of aggression, its bristled black and glossy hair seemed to hide behind two sharp curved and shiny fangs.

Before my curiosity could turn to terror, I grabbed a stick and pushed it into a big jam tin. Excited with my find, I then dug up another one, which looked just as vicious. I grabbed my trusty stick and flicked it into the same tin. In a moment it became obvious that they weren't friends, because they immediately started fighting. The fight didn't last long and only one spider lived to tell its mates.

I don't think spider fighting had ever been a sport, but it did become an entertainment that I never bothered to tell my parents about. However, when the season was right, it was something I practiced from time to time.

Years later and to my amazement, I discovered that these black fanged beasts were the deadly Funnel Web spiders. Those were lazy blissful days filled with youthful ignorance and stupidity, all combining to keep outback farm kids amused.

A nearby farm was owned by the Robinson family who also had a son named Ian, a coincidence that I felt was unusual at the time. Our farms were about three miles apart, so on alternate weekends we could be found riding our bikes to meet up. We would ramble all over the place, playing games and always skirting around the boundaries of trouble.

In 1943 and as summer approached, one weekend my friend Ian would be coming over for a visit and I remember working with timber and tools in the old shed behind the house. I had been rushing before he arrived, trying to make a small boat that I thought would be good for us to play with. Smiling to myself I could see us sailing it down one or more of the channels.

As kids we never owned a watch. However, were always aware of the time of day relying on the sun's path across the horizon to guide us. The sun timing our coming and goings, was all we needed, but I do remember this day being one very overcast and gloomy Saturday

While we beavered away, the light in the shed started fading as the farm became enveloped by a dark grey cloud. Running out of light and totally focussed on the job at hand, we did our best to complete our project.

Then, the lightning and thunder started.

Like a lot of country kids, we had been shown how to calculate how far we were away from the storm by counting the time between the lightning's flash and hearing the roll of thunder. That little shed shook as each crack from the lightning grew louder. Safe in our shelter we kept working on our project, thankful the light from the storm would show us more of our handiwork.

Our counting became shorter with each flash, until it felt as if the storm was on top of us. To say we were frightened would be over selling it. We understood the power of a thunderstorm, but having a confident level of concern kept us safe. The downpour that followed the last crack of thunder drummed on the tin roof, its cacophony filling every inch of our little shed with noise. We both understood we needed to stay put until the storm passed over us.

This morning's wheel-tracks were soon filling with clay-sodden and fast running water. We stood safe inside, looking out of the shed's windows and door, watching the deluge making a lot of little impromptu rivers and flooding the driveway. Deep with foam, this cappuccino-coloured torrent was boiling with energy. Dirt and litter from the paddocks were bubbling and tangling. The swirling mess surging its way toward the ever-swelling and nearby creek.

I looked out the small window just as the sky turned white, as it did, a frightening noise clapped overhead and instinctively it had me clamping my hands over my ears. For a moment my eyes hurt, as if I'd received a welding flash. After blinking the pain away, I waited and I just stood there staring at a changing sky. When the weather cleared, I could see that a huge gum tree that had always stood about three hundred metres from the shed had almost disappeared. Only a short and blackened stump remained, a bluish flame rising out of it. I was amazed at natures' force. The main trunk had shattered and there were still large smouldering pieces of it scattered everywhere. It was if one of Hitler's bombs had made a direct hit. That day provided our family with firewood for most of the year, but even that didn't appeasc any feeling of the loss I felt. I loved that tree.

Another time I was given six *Lane* rabbit traps, Lanes were reputed to be the best trap at the time and, I had six of them. It didn't take much pestering before Dad taught me how to set them. He showed me how to find the warrens and then select the best burrow to set the trap on. There were warrens scattered all around the farm and it wasn't long before I was catching good numbers of rabbits. I guess it was as a rabbit trapper that I tasted my first experience at being in business. I dressed and sold a few to a couple of ladies in the town and learnt how to stretch the skins on wire frames to dry. The skins were taken to Wakool where I would sell them, the effort adding to my pocket money.

From time to time we would have an unusual visitor arrive at the property. He was one of two Indian hawkers that called around the local farms selling household items such as bed linen and dress material. Their arrival was always a welcome event, because most rural women had a treadle sewing machine, which in most cases was a Singer. From the hawker they were able to select from a wide range of haberdashery items. They would use whatever they bought to make up or mend clothes, curtains and many other household and farm goods.

The Indian hawker who visited us really got along well with my dad. His name was Kader Bux, but we all knew him as Marbush. He had a covered wagon pulled by a draught horse and he would usually arrive late in the day. He'd knock on the door and ask mum if she needed any material, trying to coax her into a purchase by showing her cloth from his latest range.

We found Marbush to be a gentle man and quite religious too. I often saw him kneeling beside his wagon and praying at nightfall. He was illiterate and because they were friends, quite often he would dictate a letter for my dad to transcribe and address it to his family living in the Punjab in India. Dad told me Marbush always included money in the envelope. While we had always asked him to join us at the dinner table, he insisted on cooking for himself, and did so on an open fire near his wagon. His specialty was Johnny cake, an Indian style of bread; in today's world we know it as a wrap. He often gave some to us and dad would give him some hay or grain for his horse in exchange.

To replace any of the items he had sold from his wagon and to find new lines of goods to sell, Marbush would leave his wagon and horse with a friend near Echuca. From there, he would board the train to Melbourne and spend a week or two visiting numerous bulk stores, which were mostly located in Flinders Lane.

Another hawker who travelled the roads around our area was Besant Singh. He never visited us on the farm, but I had seen him camped within the small township of Wakool. He was quite old when I was a kid and I know he was well respected. Many locals had reached out, helping to care for him as his life started slowing down. I have since been told that he passed away and was buried in a local cemetery. I often wonder what became of Marbush, as we never heard any more of him after we moved to the farm at Blighty.

I remember often going into town with dad, the trip involved driving down a dirt track that was about two or three miles long. Huge gum trees lined the edges of the track and determined the course. Dad would be always moving the steering wheel, swerving in and out of these age-old monoliths. Offering unheard advice to the council and its road builders as we went. In later years the local council did improve the road by removing a lot of the trees and forming an unsealed, but safer dirt road in place of the track.

A "Jerry's Lane" signpost still proudly marks the council's good deed. Named after my dad Gerald O'Rourke and on the Burraboi to Wakool Road it is a warm reminder of our time there. Although there was no ribbon ceremony offered, Dad made sure he was the first person to use his new lane.

Last time I flashed past it on my way south, it pleased me to think it was still there, marking him as a legend, a pioneer, or a pain in the neck to the council that my old man was.

DAD'S ACCIDENT

In 1942, my father had been using Scotty and Bonnie to haul away a broken tree that had fallen close to our house. However, the tree struck a small stump and its trunk swung around, hitting him behind his right leg, causing an extremely bad fracture. It was nearly dusk and dad kept calling to mum and I. We raced outside to help him. Mum shot back into the house and phoned Les Tynan in Wakool. He had a new car and along with two of our neighbours came to Dad's rescue. Loading him into the car as carefully as they could, before rushing him to the Barham Hospital.

While our friends were helping dad, I busied myself unhooking our two draught horses and walking them up to the main shed. To remove the collars and harness, and me being a bit on the short side, I had to find a large drum to stand on before I could reach over their heads. Thankful that our horses were really quiet, I led them to our small paddock nearby, happy to close the gate after them as they moved away to graze.

Mister Tynan asked me, “where are the horses?”

After I’d told him what I’d done, it took a while for him to believe that I had actually managed it by myself.

Dad was in the Barham Hospital for almost six weeks before the doctors told him his leg would not heal. The break started under the knee, split along the length of the tibia and into the ankle. There was tissue growing in between the main fracture, and the Barham Hospital was unable to correct it. He suggested that Dad be admitted to the Royal Melbourne Hospital. There, doctors specializing in Dad’s condition would be better placed to treat him and repair his damaged leg.

Once again, the Wakool shopkeeper, Les Tynan stepped up, driving Dad, Mum and me to Melbourne where Dad was admitted into this huge Hospital. He remained there for almost fifteen months until his leg had healed.

While we stayed in Melbourne, the farm was being cared for by a manager.

MUM'S MELBOURNE PAPERBOY

Now in Melbourne, Mum and I moved into a suburb called Sandringham. So, for the first time in my young life, I went to Primary School where I met Eric Seccombe and made a new friend.

Somehow, I'd found a small job on Saturday afternoons and for three or four hours I'd be selling the Herald evening newspaper. Later I'd be hawking The Sporting Globe from 5.00 pm at the railway station. There were about six of us newsboys working the station and competition was fierce. As Sandringham was the end of the suburban rail line there were many people coming home from sports events, movie pictures in the CBD. While others would catch one of the many buses going further South. Nearby, there were three hotels so their patrons were good customers.

Working on commission, I would take home about four shillings, which was top money those days. One evening a fairly happy man came out of a hotel as I stood at the door. He asked me for a Sporting Globe, costing six pence, which is equivalent to five cents today. As I passed him the Globe, he gavc me a ten-shilling note, which is a dollar today and as I went to give him change, but he waved his hand saying,

“Keep it boy, I had a good day on the horses.”

That night when I took home around fifteen shillings, (one dollar fifty), boy, did I have some explaining to do. My mother, never seeing me come home so ‘rich,’ thought I must have stolen the money.

A NEW FARM, A NEW ADVENTURE

Eventually, we returned to the farm near Wakool, but my parents decided it was time to move on. It must have been sometime around 1945 when Dad bought a 500-acre irrigation property at Blighty. Our new home was now situated half way between Deniliquin and Finley in New South Wales.

Dad named our new property "RANKEELA." One afternoon, I watched him paint a sign declaring the farm was ours. When the script was dry and with my chest stuck out, I proudly helped him hang it at the entrance to our new home.

I would be attending the small school built on land that had once been part of the farm. The original owner, a man named Gregory, had back in the 1920s gifted a half-acre block in the top corner of the property to the New South Wales Education Department. Mister Gregory must have been a forward-thinking leader capable of envisioning the need for a Primary School. He could see there would be a growing number of children arriving with their parents as they moved into the district to take up their new farms.

To house us, there was an old farmhouse onto which a new kitchen had been built along one side. Close to the house stood a large shed for the farm's machinery and there was a dam nearby.

Irrigation farming was in full swing in this area as the channels necessary to carry the water had been constructed much earlier than those around Wakool. The irrigation system's water was coming from a weir on the Murray River located at Yarrawonga/Mulwhala. It flowed to our farming district via a wide canal that ran along the side of the main road, then down through Berrigan, Finley and on to Deniliquin. In later years the water was piped under the Edward River at Deniliquin where it carried onto Womboota.

These two new areas were named the Berriquin and Deniboota Irrigation schemes.

A huge construction and naturally called the Mulwhala Canal, it was at least twenty metres wide and probably at least two metres deep. I remember throwing a stick into the water and watching it disappear, not a torrent, but it was flowing fairly fast. I was told that the flow rate was 1,400 cusecs (cubic foot) per second. It had been suggested that at the time, this was the largest man-made canal in the Southern Hemisphere.

A couple of years after we arrived at Blighty, the Government extended a smaller version of the canal via a branch that went past our property and then northwards towards Mayrung.

Construction was done with three massive Buycyrus Draglines, machines we know today as excavators. The first machine would drag up the top soil, placing it way out to each side. Then the other two dug deeper into the centre, placing the subsoil onto each edge and forming banks known as batters. As they worked, these huge machines rolled over each bank, compressing the large mounds and compacting the soil.

By today's standards, these excavators are only small. However, at the time each bucket held ten cubic metres of dirt and one day to illustrate their size, an operator manoeuvred his car into one of them. His demonstration showed there was enough room for a person to easily walk around between the car and the sides of the bucket. Watching all the locals standing aghast and being unable to comprehend its size of it is something I've never forgotten. This simple act would become a memory that served me well in later years. Something I could rely on when attending field days and presenting the attributes of different tractors or machinery.

During my early school years, I learned that in order to give returned servicemen a new start in life and acclimatise them to normal times, properties of similar size in our area were allotted to soldiers returning home from World War I. This explained in part the number of small holdings, which were referred to as, Soldier Settlements.

One of our first jobs on the farm was to build a small milking shed to service the milking cows Dad bought to provide an income stream. He also sourced some breeding stock to ensure we had new cows to come on after the milkers were turned off.

DENILIQUIN HIGH, HERE I COME

In early September 1945, arrangements were needed for me to finalise my Sixth-Grade schooling. My parents believed that to make it easier for me to transition into high school in the following year, it would be best to attend the bigger school in Deniliquin. This meant I would now stop attending the little school at the end of the property and begin studies at the Deniliquin Intermediate High School. Its name being such a mouthful, everyone had shortened it to DIHS.

We were living about twenty-five miles from Deniliquin and the school bus ran from the Depot at the Blighty Hotel. A centre for the community, it was affectionately known as the MUD HUT, because it too had also been constructed of a pizzey-mud potion. The same material that Dad had used to build our home on the Wakool property.

Because getting to school by catching the daily bus run would be quite arduous for me, my parents decided the distance made it impossible. It was because of this, they arranged for me to stay at the hostel located near the local airport.

I would be there for the five school days and have my weekends at home. During the war, the Government had constructed an airport with barracks large enough to billet five-thousand air force personnel. As the war had ended in 1945, all of the accommodation huts were sold or moved onto the new soldier-settler farms, leaving only the RAAF hospital on site.

Just as Mister Gregory had gifted land to the education department years earlier, a group of enterprising locals had banded together and purchased the hospital. Their vision was to offer children living on farms too far from the end of the bus runs, somewhere to stay. The end result was a four dormitory complex capable of accommodating sixty children. These were farm kids who attended a variety of classes from grade one, right up to year ten. It was staffed with five adults from the town and four school teachers who would supervise us for the four evenings we were there.

The first day I attended was at the start of the last term of Sixth Grade. There I met another boy, also on his first day. We were the same age and were told to sit together in the last desk in the very back corner of the class. Back then, all of the seating was decided by the teacher. They placed the students in order from the front desk to the back row in accordance with their exam marks. Who would be our smartest classmates was a fact not lost on us? The twin Kirkpatrick sisters had been occupying the front desk since Grade One.

On the first Friday, a full exam was held on all subjects. My new friend Brian and I felt happy about the answers we had given to the questions, especially when the teacher announced the results on the following Monday. I'll never forget the pride filling my chest that morning when Brian and I were moved to the front desk. The twins now occupied the second. Wow! who could have imagined that?

Each Monday morning, five other children and I would ride to school in a local farmer's ute. Then on Friday afternoon, another parent would pick us up and return us home. This happened on a weekly basis and our parents would pay a small amount to the drivers to carry each child.

For me the saddest part during that last term was Brian Teelow's father passing away and him not returning to finish school at Deniliquin. After his father's death he and his mother moved back to Thargomindah, in Queensland. However, I would get to meet him again in later years in a surprise situation.

School lunchtime consisted of a round of sandwiches, plus a small bottle of milk. Lunch had always been prepared by a local ladies' group who were running the canteen. On many occasions boiled egg sandwiches were the base meal. Needless to say, for many years after, I could not eat hard boiled eggs of any fashion.

After two years, many of the school bus runs were increased in length. Now that the bus stopped on the main road just over a mile down a lane from our house, it made schooling much more accessible to me.

Around that time, my father had another bad accident while digging a post hole. The crowbar he was using had skewed sideways, tearing its way down his right leg. The same leg he had hurt at Wakool. This meant he had to spend quite some time in the Deniliquin Hospital until it healed. It must have been bad because when he came home, he couldn't walk very far. His situation posed a substantial dilemma for our small family. Dad had been milking nine cows by hand every morning and using a hand operated separator for separating our milk from the cream. The cream would be decanted into churns and left for the truck from the dairy factory to pick up. Any residual milk we would feed to our poddy calves and a couple of pigs.

Plans had to change. Now I had to get out of bed even earlier. I had to step up and help my mother manage everything Dad had been doing. By then it was time to have a quick bite of breakfast, change into my school clothes and ride down the lane to catch the bus at 7.50 am. I would arrive home at 5.00 pm and then, we would start the process all over again, have dinner, do my homework and try to get some sleep. This continued for ten weeks and while the farm routine remained the same day in day out, school only lasted five days. So, my weekends were much better and something I could look forward to.

Time was moving on and in December 1948, I graduated from school and proudly accepted my award, the "Intermediate Certificate." At the time this was the highest qualification available at DIHS

With school behind me, I was now working full time on our property. We had built a three-stand milking shed and installed "Royal Medal" milking machines to cope with more cows that were due to come on board. As many people know when milking cows, it is necessary to have enough ongoing feed arrangements in place to keep the milk flowing. With irrigation available, the job of a dairyman can become easier or harder, depending on which side of the fence you are sitting on.

Naturally, irrigation does not work for eight hours each day, so the farmer has to keep checking the progress of the water as it floods over each bay. When those bays are nearly finished flooding, the slow flowing water must be directed to new bays. It's a grinding routine that can wear one down.

One moonlight night, around 10.00pm it was time to change the gates on the finished bays and open the dry ones. This was done by shovelling dirt into the gate, which is no more than a gap in a minor channel bank, then with the shovel, cut a new gate through the bank of a dry bay to flood it. So, with the shovel over my shoulder, I mounted our pony Frisky, and riding bareback, cantered up to the channel. I jumped off, leaving Frisky to graze on some fresh lucerne, while I filled in some of the dirt openings, then moved on to open up fresh bays. Suddenly, the moon went behind some heavy cloud and as it did, everything went black. To add to my troubles, Frisky had decided to head home by herself. Frankly, I had a big problem, I couldn't see our house light, nor a car on a road, so it meant me walking slowly until I reached a fence, carefully feeling for the barb wire then the netting below. Knowing every inch of our fencing, it was obvious to me that I had to walk left. It was slightly downhill and I'd walked for a few hundred metres when low and behold my spirits lifted, as the house lights started coming into view. Walking into the back door I was greeted by my frantic mother. She had become upset when Frisky arrived home without me. It didn't take much of an explanation from me, and before soon, all her fears were settled.

Turning sixteen was a defining moment in a farm boy's life and on the morning of my birthday, I presented myself to Sergeant Pickering at the Deniliquin Police Station. I was there to gain my driver's licence.

Looking past me and through the window, he politely asked, "How did you get here?

I responded, "I drove dad's truck in from home."

Pointing his pen toward the window, he said, "when I saw you drive up, I thought you'd be coming in for your licence." He typed out the form, then asked me for two shillings and sixpence. (Twenty-five cents in today's currency) Passing me the licence he added, "next time you come into town, bring in a butt of wheat for my chooks."

During summer that year, I joined the Blighty Cricket Club. We played in a local district competition against Mayrung, Retreat, and Finley. Each game played out over two

Saturday's. My batting was pretty hopeless, but I really liked leg spin bowling. There were no scheduled practice days, so during the week, each person had to make do with practicing by themselves at home. My dedication to spin paid off by becoming a more than handy leg-spinner.

In mid-1949, my father presented me with a new Cooper Handpiece. This is the power-driven tool that shearers use to remove the fleece from sheep.

Handing it over, he said, "I have booked you into a shearing school at Echuca.

There, I learned how to shear, which was great as it didn't take long for me to get the hang of it. After I became competent, I teamed up with another "learner" in the winter of 1950. The milking cows were 'dry' and for ten weeks we went shearing. We were working in different farmer's sheds in and around the local area. The going rate for shearing was seven pounds and fifteen shillings for one hundred sheep per day. Shearing was earning me about forty-pounds each week, which was terrific money during those days. I kept shearing for three years.

In the early fifties, I had been milking up to sixty cows twice a day and for a period of nine-months each year. During that time, I had watched as many of the local dairy farmers had started fitting 32 Volt lighting plants into their homes and sheds. Being interested in anything mechanical, I was more than curious to learn how these power plants worked. I wasn't envious, but I could see how having electricity on our farm would improve our lives.

After I had studied various publications on 32-volt power and having bought a new Holden ute, drove to Melbourne. There I visited several electrical warehouses, filling the ute with sixteen 2-volt batteries, numerous switches, a 32-volt generator and enough wiring to electrify our home. Driving back home I kept running through all the things I'd heard from the experts I had met. I started to marry some of their suggestions into my original plan and by the time I reached the farm had the complete design figured out.

A year or two earlier, we had erected a water tank on nine sturdy poles near the old house. We had positioned a three horsepower Ronaldson Tippet petrol engine on concrete foundations underneath it. Coupled to a pump which drew water from a dam fed by the channel, our engine would fill the tank and from there, gravity would feed a series of pipes supplying water to the house, milking shed, and piggery.

My plan for power, meant coupling the engine to the new generator and running wires to charge the batteries. Within a week I had the unit running and the house fully wired. Bingo, we now had wonderful lighting for the first time. To say we were all happy and relieved to never light our old smelly kerosene lamps again, was an understatement.

Still shearing when not milking, my next big purchase came in the form of a diesel-powered Nuffield tractor. At the time, it was the only brand available. We used it to sow part of the farm to barley, oats, and wheat. In addition to the cereal crops, we grew lucerne, which a local contractor baled for us.

An official government letter addressed to me came in 1956. It stated that I had been summoned to attend Puckapunyal Military Camp, located near Seymour in Victoria. There, I would be serving my country with ninety-eight days of National Service military training.

National Service 1956
Ian O'Rourke second from left on the second back row

It put a kink in a few of my farming plans, but it the military calls, you are not in a position to argue.

TIME TO LEAVE THE NEST

Work was continuing on the farm, but after my army training, I had become restless. I felt that I wanted to do more with my life than spend it milking cows, shearing and farming. There had to be more out in the world for me. Moving away would be a new beginning and after getting my thoughts together, I gathered enough courage to tell my parents that I wanted to try and find a job in the city. Still wanting to be working in the Agricultural Industry, I told them I would try to find work at one of the many machinery businesses based in Melbourne. While we talked it over, I noticed a flicker of sadness in my mother's eyes, but both of my parents encouraged me to give it a go.

In January 1959, I drove to the big smoke, Melbourne and began looking for work. I set my sights on a position with one of a number of companies based in the Braybrook area to the west of Melbourne. My first stop was at Massey Ferguson in Sunshine. There I spoke to the Personnel Officer and explained what I was wanting to do. After a number of questions that felt more like an interrogation than interview, he told me Massey were planning to open a new Department. This department would be charged with testing recently designed farm machinery. Saying he thought my skills would be valuable to that team, he sent me around to an office where I met Jock Berry. Jock an experienced machinery man, had been appointed to the newly created position of Chief Test Engineer.

After several minutes and talking as you would when meeting a new friend, Jock said, "I think I can use you, but I haven't been given a budget, or my staffing levels yet. So, if you'll agree to working on the assembly line until then, I will call you back in a week or so."

Next day, I started work on the assembly line assisting a fitter who was building the seed and fertilizer distributor components for the MF562 Sod-Seeder.

He was building ten units per day, but kept dismantling each second or third finished section, then rebuilding it. He claimed each one was not totally correct. Frankly I was amazed, as every part seemed to be working fine when he ran his tests. At the end of the first week, I was told he was going on leave and I was to do his job. I built over twenty units on the first day and was surprised when at the end of work, I copped a dressing down by the Union Rep for going too fast.

Thankfully, Jock Berry came down to the factory to see me and said, "come on, my budget has been confirmed and you're working with me now."

When I went to the test shop, I met an engineer who was Jock's assistant, but the department was growing and in six months there were seventeen of us working together. Two of them were a father and his son, several mechanics, four mechanical engineers, four more men with different degrees, and five keen workers like me.

Ian Testing Massey Ferguson's prototype MF 585 Header

My first task was learning about the new MF 585 SP Header, a self-propelled combine harvester. Once confident, I was tasked with conducting reliability testing by driving over a set of steel pipes of six inches, or one hundred and fifty millimetres in diameter. Speed over the track was changed after a set number of passes. Examination of the header was checked at the end of specified periods. The 585 had some new and interesting design features, such as the ground speed variator. This mechanical device was basically three large V pulleys with the two end sheaves being in a fixed position, while the middle pulley could be moved by a lever on the operator's platform. This would change the forward or rear speed operation at any given time. Quite an ingenious invention and reinforced my thirst for understanding the working of machines in this emerging mechanical world.

Whereas the North Americans and Europeans called a harvester a combine. In my early days, a combine was a seed drill combining seed and fertilizer when sowing. These Australian designed machines employed a ground driven metering system. Once through the metering cogs, gravity would draw seed and fertilizer down rubber tubing running down the back of the sowing tyne until it settled onto the seedbed where harrows following the drill would cover it with soil.

It is an interesting point that the name Header was used in Australia to denote machines that would cut off the heads of cereal crops, such as barley, oats or wheat. Also, the front of the header had a closed, or comb front. Later in time, new products coming in from overseas were called combines. These machines were usually fitted with “open fronts” or “flat table fronts,” which enabled the machine to pick up crops that were flattened by storms. These bigger machines could thresh much greater quantities of the bulk product than earlier Australian designed units.

I had no difficulty working with Jock and because I was the only member of the team who had any farming experience, he often asked my advice. When we had to do field tests, Jock would often question me if it was dry enough to test machines in cultivation, or if it was right to go mowing or baling hay. I soon started learning that knowledge I took for granted had worth to those who didn’t understand it, and it felt good to feel valued.

During this time, I lived with four men in a house near the plant and I always arrived at work early at 7.00 am. Being in first, Jock asked me if each day, I would update test results on the large blackboard mounted to one of the walls. These outcomes would be taken from each operator’s job sheet, something they had completed at the end of the previous day. Having completed this each morning, I would watch Jock standing and studying the board, obtaining a quick idea on each project’s progress. It may have been a simple request, but collating those records did save us a lot of time.

During my time with the Test group, I had made friends with two of the guys, Stuart Curnow and Ian Donald. Along with school teacher John Millhouse, we decided to rent an apartment in St. Kilda. There we were joined by another chap who was an engineer with Mobil Oil refinery at Altona. All mechanically minded, thirsty for knowledge and eager for what came next, my time sharing with them was a ball.

At Massey Ferguson, as each newly designed machine was passed over to the Test Team, Jock along with the Design Engineer would have a discussion on the elements of the testing procedure and create a series of trials. Then Jock would allocate one of the team to do the job and report their findings.

As each machine we received was a new design, our team would take time to study the design objectives and explore the operation intent. Usually this was completed in the Test Workshop, after which the designated testing operator and Jock would discuss the next steps for putting the product to work in the field.

Each week our whole group would attend meetings to discuss the progress of each product, whereby fresh ideas would often arise and ascertain if all was proceeding to plan. The machine's designer would often join our meeting and add his thoughts to the discussions and make sure that we were running tests in line with his original design concept. As could be expected, faults may have occurred that would need addressing by the design team. However, most of these items could be repaired quite quickly next door, in the Engineering Workshop.

Following my initial project test of the MF585 SP header, I completed my report, then would sit down with Jock Berry to cover any areas that needed to be changed. In North America, self-propelled harvesters had been around for decades, but where Massey Ferguson's 585 SP excelled was its simplicity. Its ease of manufacture, ease of operation, simplicity to repair, and reasonable cost. In these areas, the 585 proved to be a worthy product, and probably one of the earliest headers in Australia that were completely self-propelled, replacing the horse or tractor drawn HST McKay units.

My next project was to test the MF 585PTO header. As usual the prototype was towed into the test shop for initial checks and stationary operation, all of which worked well.

Additional tests were completed at the test track in the main factory yard. There, we would be checking for any possible structural failures. The prototype 585 was hitched to a MF65 tractor, first in an unladen state then the grain bin was filled with a set amount of sand for further action. As a group, we were pleased when the machine passed this testing stage with no need for any design changes.

My step was to conduct field trials in a wheat crop located on a property near Melton, again the prototype 585 PTO header worked extremely well. However, Jock and the designer were slightly concerned over the lifting speed of the comb, as they believed it was not quick enough.

The machine was brought back into our test area to carry out a speed-trial. I was not present when the unit was first set up, but arrived to see Jock standing by the 585, where he had the tractor engine running at 2,000 rpm.

Jock said to me, "Get up on the tractor and drop the front."

Once in the seat, I moved the remote hydraulic lever on the tractor to lower the header front. Within one second there was a tragic noise as the whole front comb buckled into a U shape.

Needless to say, there was a stony silence only broken by the murmur of the boys in the shop surveying the damage. The header front had not been placed in the correct operating position. Instead, the comb was in the road travelling position, a feature that made the drawn machine narrower. As a result, the comb hit the 585's towbar on the way down. Although an easy thing to overlook, this was a complete disaster!

I assured Jock it was my fault for not checking the heavy towbar's position prior to getting on the tractor.

He rejected my comment saying, "I got the unit ready myself and then asked you to get on the tractor to operate the lift. Ian, this is not your fault."

The next step was to go upstairs to see the Chief Engineer, J.K. Gaunt. A man who did not take bad news well.

Jock started the conversation rather bluntly by saying, "we have just totally wrecked the 585 PTO header,"

Mister Gaunt threw a few expletives at Jock before asking again, "totally wrecked?"

Jock replied, "well the drum and concave sections are only slightly damaged and can be repaired."

Gaunt's reply was only a little more upswing, then Jock said, "well the straw walkers and housing are quite okay."

That brought out a sigh of relief from the Chief who told Jock and I to go back and get it fixed.

On the way back down the stairs, Jock gave me my first lesson in how to decrease a major disaster by going in heavy at first, then reducing the drama by taking the big log off the burning fire. Hallelujah, we had minimised what could have been a major concern for the careers of all the test team.

The moment we arrived back at the wreck, Jock told the workshop to, "fix it," adding, "weld a metal stop on the towbar so the comb cannot be lowered while it's in the travelling position."

Jock having given his instructions felt our work had been completed. We had unearthed a problem that would have surely happened again in the field. Our system was working. Job done.

When repairing the machine, the workshop team welded a new stop into position and the 585 was ready to continue testing. The role of a test team is to try and break the machinery. It's part of their remit and bending that header front on the MF 585 PTO was proof that the initial design required more development.

Although it seemed like a disaster at the time. Without the stop being fitted, it was easy to envision warranty claims flooding in from the field because owners and operators had forgotten to take the machine out of its travel position, before lowering the header front.

MF 585 HEADERS and DRUNKEN PIPERS

In 1963 I married and bought a house in Box Hill North. However, the union was destined to fail and after twenty years was dissolved in 1983.

During my time in Box Hill, I was approached with an offer to join the local School Council, which I did, and when asked, I found myself suggesting various ideas for improving several things around the school.

One item raised my attention; the playground had an unmanageable slope that grew hardly any grass on it. To my mind, this presented two major problems, in wet weather it would be impossible for students to play on without slipping over, or suffering injuries.

I put a plan to the School Council that we should run a big raffle and raise funds to construct a better playground. Frankly, the opposition to the project was surprising but to overcome the curly questions, I had another idea.

Each of the female councillors would go through the students records and gain the address of all the parents. At the same time, I would be approaching outside parties to donate free prizes. Qantas agreed to offer a double return ticket to Singapore. Canberra Television donated a free reverse cycle Air Conditioner. This came from Fred O'Neill, a fellow councillor who was CEO of Canberra TV. We managed to get a third free prize from a local shop owner.

I wrote to the Victorian Lotteries Commission for approval, which was granted. On behalf of the school I also approached Box Hill City Council for permission to use their Earth moving equipment to reduce the grade to a much gentler slope. One Saturday and because they were fathers of the school children, we had several council drivers operating their equipment free of charge.

Another of the parents had a printing business, so he printed and bound free raffle books of ten tickets. A raffle book was sent home with each child to sell at $2.00 per ticket, the proceeds to be returned to an appointed Treasurer.

Within a month, the majority of tickets were sold with our funds swelling to $2,300.

Occasionally on a Saturday, the school council would run a small fund-raising effort to add extra coffers to the school treasury. On the designated day popular television personality Bert Newton was gracious enough to draw the winners of our raffle.

It was a complete success and every school councillor agreed that without the help of the Box Hill Council and the operators of the equipment who did a great job on that Saturday, the kids would still be playing in mud. The raffle had raised enough money that we were able to pay a landscaper who covered the whole area with hardy lawn.

There was a sequel in the following year. We found that the grounds needed regular watering and drainage. After a heavy rain, something was required to take the water away. Once again, I fronted up and ran another raffle. This time it raised about $2,000, which was enough to finish the playground.

Back at Massey Ferguson, we had a new set of machines to be trialled. These were part of a new haymaking range, a 503 baler, two cutter-bar mowers and a side-delivery hay-rake.

After the normal check over in the Test Workshop, arrangements were made for the machines to be transported to a property at Warragul in Victoria. It was owned by Walsh brothers. They were haymaking contractors and really great guys to work with. They had plenty of contracting work to do and having our team with them, we could get through a pile of tests even though our team had to stop at times to review the results. We usually worked on weekends and were staying at the main hotel in Warragul. We lived and worked there for over a month.

Arriving back at the Sunshine Plant after the Warragul assignment, I was given a task of test fitting many different brands of three-point-linkage machines onto our MF range of tractors. My role was to ensure that all of the machinery presented could be fitted with ease. For those machines that wouldn't hitch up, I would report my findings and make several suggestions and if I thought they needed a kit to achieve the right result, I would rough out a sketch for the engineers to follow.

The next job was trialling the Massey Ferguson 585SP Header in rice. We took it to an area near Griffith, New South Wales. This combine was fitted with a new design 'Open Front' that was intended to strip rice when it was laying down on the soggy ground. I found this a difficult task as the rice fields were so wet, that it made it near on impossible to uncouple the front at times. Removing the front was a necessary step in allowing us to move the combine back. A move necessary to inspect the feed pattern going into the drum and its concave. Once again there would be photos taken and a report sheet to complete that explained them.

My troubles started when I began trying to recouple the two sections. The combine, due to the muddy conditions, would slip, never returning to the same angle. This would leave me working in mud and slush until finally getting the two pieces to meet, I would then have to make sure they were fixed together correctly.

Frankly, the quantity of material coming into the machine appeared to overload the original design concept, but the observations that I'd entered on my field report were dismissed.

Much later, I accepted a promotion to work as the Service Representative for Western Victoria. I began visiting new MF 585s in the field and found I was responding to owners complaining about feed issues when they were working in downed wheat crops. These were the same issues I had reported when testing the MF585 in rice crops. It vindicated my earlier assumptions and I continued to refer to my early findings when reporting feeder faults to engineering.

While living in Box Hill North I was approached by a neighbour, Jack Harris who was involved with a few of the Scottish pipe bands in Victoria.

He asked if I could be marshal for their events. Following my stunned look, he told me the Scots were looking for four guys to assist the bands when they had competitions in Melbourne. These marshals would usher each band into position and get them there at the right time to enter the grounds and perform. In other words, my job, along with the other three, would be to just keep things together and on time.

I joined up with Jack and two other gents, and in good Scottish tradition, we purchased our green jackets. These were embroidered with suitable logos. On the appointed Saturday or Sunday, we would meet at the location and fulfil our duties. After the first couple of meetings, they let us know that our duties were well received by all concerned.

Our biggest challenge was a huge promotion to be held at the South Melbourne Football Grounds and to be conducted over a very long Sunday. As expected, the whole event turned out to be well attended by many spectators and bands from all over the country. all went well through the morning, then lunch and into the afternoon.

However, late in the afternoon we had to line up all the bands. They were to enter through a small gate and arrive onto the grassy football field. Now most of us know, that over a celebratory lunch, true Scotsmen like to enjoy a little mateship and rejoice in their Scottishness with a little tipple. Well, getting them all to stand up, their pipes at hand, assembled into their correct groups and then into lines, made our little band of marshals busier than ever. Picture in your mind, trying to gather together all of these more than happy guys all pumping out "Amazing Grace," then leading them, six men abreast, through a gate no more than eight feet wide, but like true Scots, they managed it

Thirty odd bands finally filling the stadium was wonderful, the ground was heaving, filled with enormous sounds and such incredible colours, it made my heart sing. Those glorious mass bands all beating in time with sounds and colour is something I have never forgotten.

A DESK JOB IS NOT FOR ME

My next promotion came before I had finished the rice test at Griffith.

Jock called me back to Sunshine where he said, "I have some interesting news; I have been promoted upstairs and I would like you to come with me. It will mean a pay rise and your new role will be to write test plans and work with the designers on any new projects."

It was a good deal and one I couldn't pass up. So, we shook hands and I moved upstairs and into nice surroundings. I had plenty on my desk to deal with, but it was a bit heart breaking when I lifted up my eyes and took a new glimpse of the scenery. At any time of day, all I would be looking at were the rusty roofs of industry. A constant stream of blue-black smoke from Sunshine's factories filling the air drifting over them.

After two weeks, I had to tell Jock that I liked the job, but the outlook and being tied to a desk was getting under my skin. He tried to get me to stay. However, my decision had been made. Jock told me he thought I was a valuable asset to the company and in an effort to keep me phoned the manager for Massey Ferguson, Victoria. That office was based in another company building on the Melbourne to Ballarat Road, five hundred metres away.

After Jock had made his call and his discussion with Cliff Freer, Cliff took me on board as the Service Representative. A role that made me responsible for the service needs of the company's dealers in that area. My territory, would be Western Victoria and it covered an area that ran north from Sunshine to Kerang, west to the border of South Australia, and down to the Southern coast. In all I had eighteen Dealers to visit. It was some territory to cover, but I must say I was well received and found them a friendly bunch of people.

MASSEY FERGUSON'S SERVICE REP

Not long after joining Melbourne Branch, as it was known to staff and dealers, Cliff Freer left the role and was replaced by Derek Johnson. A man who in my mind believed that after sales service was an unnecessary evil, and by his nature, service personnel were made to feel they were not welcome in his office. Referring to him as not one of my favourite people, is an understatement.

Massey MF 585 Training at Hamilton Victoria

Therefore, I was pleased when I was introduced to my new superior, Ron Sheldrick. He had been installed as Branch Service Manager. Ron proved to be much more than a boss, he had a great attitude and someone I could easily report to, but his worth to the company lay in his experience. He had an innate grasp of anything mechanical and being a great problem solver, I found him to be a deep font of Massey Ferguson knowledge.

My new role included a company car and a shiny new Ford panel van became my assigned vehicle. My appointment came around the same time that the new MF 585 PTO header was released to the dealer network. Twenty-five had been ordered and shipped to dealers in my new area. The first problem I encountered was that the company had failed to ship Operator Manuals with the machines. Dealers were jamming the company's switch board with complaints.

An operator's manual is an essential part of the machine. It is a vital reference for the initial setting up of the harvester so it can reap differing crops, and offer a raft of alternate settings for different conditions. The situation only exacerbated when dealers became aware that Head Office had fallen behind in writing the manual prior to the main release.

Needless to say, this created some concerns with the dealers, but it didn't bother me as much.

Due to my experience with the field-testing programme and my knowledge gained in the Test Department, I could refer to my copious notes and lean back on my general experience. Now faced with a lack of support in the way of manuals and dealers hungry for knowledge, I spent those first few weeks covering many miles over highways and rough dirt roads passing on information. By the end of the season, I was proud that because of my efforts, I had helped to train Massey Ferguson's dealers and their customers in the correct procedures for operating their new MF-585 PTO headers.

During the harvesting period and because my territory was so large, I spent long hours driving, often travelling about fifteen hundred miles per week. The driving between towns often happened after the dealership closed, which meant I was working long days. It was often well into the night before I could check into my hotel. The pace was wearing.

However, relief was coming. One day late in 1962, I called into the MF Dealer in Woomelang, it was a normal visit and as I walked in a sprightly chap greeted me and said,

"G'day Ian, my name is Noel Howard, I'm new to Massey Ferguson and as from today, have been given the top half of your territory."

I immediately warmed to this personable young man with eager eyes and, as Noel was unfamiliar with the new harvesters, that first meeting included a short rundown on the operation and servicing of the MF585. He asked questions and took copious notes, keen to soak in everything I had shown him.

What a relief, and as my area was so large it became difficult to address an increasing number of concerns being raised by dealers. The area being so spread out and with long distances between dealer points, it often required excessive time to travel between the wounded machines. Elation overwhelmed me, I now had someone to share the service load and as a result of that original meeting, Noel and I remained firm friends and work colleagues for over fifty-nine years.

On a normal visit to Rupanyup and speaking to our dealer, Jack Emmet. He introduced me to one of his valued customers, who during the previous three months had purchased an MF 65 tractor, a seed drill, and a 585 PTO header. As it was around midday, Jack suggested we all go next door for lunch at the local Hotel. During our hour-long chat, we covered in much detail many aspects of each of the customer's new acquisitions.

I knew Jack had set me up and I paid the princely sum of six shillings for the customers lunch, and as was company policy, no alcohol. At the office on the following Monday, I handed in my expense report including the hand written docket with the hotel's stamp imprint.

Within a few short minutes the wrinkled and creased expense report came back to me as if it were burning a hole in Derek Johnson's sweaty hand. He slapped it down on the desk, looked at me and with a savage sweep of his black ballpoint, scraped two lines across my six-shilling docket and in a very loud voice, said.

"You know the rules, Ian. Service people do not have authority to spend on goodwill."

Anger rose inside me, but I said nothing.

During the following week I obtained a "*genuine*" receipt for a small vehicle repair costing ten shillings. It passed the acid test and with no rebuffs.

Back to work on the road and farmers using the new MF 585PTO machines were reporting a service fault in the threshing assembly. After inspecting several machines owned by farmers reporting the same fault in my area. I found that the concave assembly was bending in the centre and engineering had found the root of this problem could be traced back to incorrect steel being used in production.

To correct this situation all of our service guys were given a "Go/No-Go gauge," accompanied by a Technical Bulletin and an instruction that required us to visit each machine and place the gauge into the middle of the concave to ascertain if it should be replaced.

Visiting one farm in the Mallee, a Dealer Serviceman accompanied me to a property where the MF585 PTO was standing outside the shearing shed. We went into the shed and met the owner, Keith Geneale. He and a mate were shearing some old first cross ewes. His co-worker appeared to be having a fair bit of trouble with his ewe and I suggested to the owner, that bloke seemed to be making heavy weather of the job.

I copped a stiff reply, “you city blokes come out here and think you know it all. I’ll bet you couldn’t do any better.”

“I don’t mind a challenge,” I answered.

The owner wiped the back of his paw across his sweat soaked brow and said, “I’ll bet you two quid.”

Having accepted, I dragged my victim out of the pen and across to the shearing stand.

“You picked a good one,” he said.

I winked and said, “beginner’s luck,” as I picked up the handpiece. Four good strokes down the belly and I threw the wool into the owner’s chest.

He turned and started to walk out of the shed "I'm going to get the money," he said, "you've done this before."

After our first encounter the tension had now dissipated and we went over to check his machine. Finding everything in order I said no to taking his money, shook hands and left the farm. It was quite unusual that whenever I visited that Dealer again, the same owner seemed to sense I was in the district and he would arrive to ask questions about his machinery. However, after shearing that ewe, he never seemed to challenge my answers.

The long milcs, and finding out that I still had a bigger territory than another colleague, and that he was receiving three pounds more salary than the paltry twenty-five pounds per week Massey Ferguson were paying me, was another thing that was starting to get under my skin. To rub salt into my wounds, my colleague only had five Dealers to visit, and they were all close to his home.

At all the regular company meetings, we Service Reps stood back and watched as most of the Sales guys were handed various gifts for outstanding achievements. This was just one more thing adding to my list of grievances. Although vital to the image of the company, we servicemen were always in the background and lucky to be allowed maybe five free beers before these meetings closed. The mood of the whole place seemed skewed and off kilter under Derek Johnson's stewardship.

A FORD FUTURE FOR CHRISTMAS

Being on our Christmas break at the end of 1963, and because I wasn't looking forward to returning to Massey Ferguson, I had been regularly scanning the situations vacant columns of the paper each Saturday. Then early in January 1964, I couldn't help but notice the bold column of advertisements in the Melbourne Herald. Each advertisement explained the Ford Motor Company was currently developing a new product range and, their division in Australia was seeking to employ people with tractor and machinery experience. They were looking to staff all areas including sales and service positions.

Fresh from my run in with Dereck Johnson and others in the Massey Ferguson hierarchy, I wasted no time in forwarding an application, which resulted in a prompt reply that included a date and time they expected me to attend for an interview. On the allotted day and at the time they asked, I made myself known to Ford's Human Resources personnel who were based in SHELL House, Melbourne.

Two days after my first interview, I received a telegram from Jack Walker, Service Manager of the new Tractor Division, inviting me to meet him in the Tractor premises in Coburg, on Friday 7th February, 1964.

I ANSWERED THIS ADVERT AND JOINED FORD ON 10TH FEB 1964

Page 34—The Herald, Wed., Dec. 11, 1963

EMPLOYMENT EMPLOY

Ford

TRACTOR and IMPLEMENT DIVISION

REQUIRES

PRODUCT TRAINERS

To increase the spread and depth of information on service and operation of our "Fordson" Tractors and associated Implements we are creating several new positions in our Product Training School.

Those trainees appointed will be men whose product knowledge is both practical and up to date and who are experienced in passing this knowledge to training groups by lecture and demonstration.

At our Tractor and Implement Division, Head Office, North Coburg, the trainers will be engaged on the preparation and delivery of courses of instruction.

Additionally, they will be capable of conducting and controlling groups/lecture and demonstrations/in product training away from the Head Office. This will entail regular short visits to country areas both in Victoria and Interstate.

QUALITY PROCESSOR

A sound mechanical background is required together with an extensive knowledge of the Product Range.

The appointee will be concerned with the quality of incoming material and field operational problems, the rectification of any damage or transit claims and aspects of technical service.

These positions offer worthwhile career entry prospects to ambitious men who are looking for an entry to our expanding Tractor and Implement Division.

Written applications, marked with the position name should be posted by December 16, to:—

Salaried Employment Officer:

FORD MOTOR COMPANY OF AUSTRALIA LTD.

155 WILLIAM STREET, MELBOURNE, C.1.

TERTIARY STAFF REQUIRED

COOK

DENTAL NURSE

ENGINEERING

HOSPITALS

KELVINATOR AUSTRALIA LIMITED

Accountancy Student

MARINE INSURANCE

JUNIOR (MALE)

It only took us a few minutes before an agreement had been reached and I felt on top of the moon. Jack asked me what salary would be acceptable. Drawing up to full height, I suggested thirty pounds a week seemed fair.

Jack must have seen some promise in me and said, "Would thirty-five pounds, ten shillings be more suitable?" adding, "be here at eight o'clock on Monday morning."

Walking me to the door of his office, he told me they had received seventy-five applicants for six vacancies. I felt my shoulders lift, knowing that he valued me for my experience and knowledge. Feeling valued was the very thing that I knew was missing under Derek Johnson's employ.

My big move was scheduled to happen on Monday, 10th February, 1964. Now it was time to visit Massey Ferguson for the last time.

On my arrival at Massey Ferguson's Ballarat Road office, Ron Sheldrick greeted me asking, "Have you come to pick up your van?" Massey Ferguson's rules prevented service staff keeping their vehicles over holidays.

I handed Ron my letter of resignation, he looked strange and without saying anything, rose out of his chair and entered Derek Johnson's office. On his return he said, "Derek wants to see you now."

Entering Johnson's office for the first time, I waited for him to finish writing something on a memo pad.

His manner annoyed me. I had been a faithful servant for Massey Ferguson, and under Johnson's management, the company had been treating me like rubbish. I wasn't holding back now. "No, not a paltry pound at all," I said, "It's more like a fifty percent increase."

He put his hands on his desk and pushed up. Leaning forward, his face started flushing to cherry red, and with spittle creaming in the corners of his mouth said, "that's rubbish Ian... Believe me, I'll personally see to it that you never get another job in the tractor and machinery industry."

I turned on my heel and walked out. From there I went back to see Ron Sheldrick. We shook hands and said our goodbyes. I liked Ron, but had no remorse in leaving him to the toxic company of the many puffed up, insufferable, and angry men who inhabited that building.

Arriving at Ford's tractor operations to begin my new job, I was surprised to see two other men from Massey Ferguson in the new line-up. Ron Somerville and John Vaughan had also agreed to come onboard. They were with Col Knox, a technician who had joined from New Holland.

Another Massey Ferguson colleague, Peter Newman was already on the payroll, having joined Ford some months earlier.

About two months passed before Roy Pinney and Les Ward joined our team, expanding it to six Product Trainers. We all had different experience with a wide range of farming machinery, which meant we complimented each other's skills. In those first few weeks we gelled to make a tight knit service team and the cross pollination of our various technical abilities made us a formidable team. Through Ford's internal grapevine, we discovered that someone high in Massey Ferguson's management had phoned asking for Ford's General Manager, complaining that Ford's tractor division was poaching their experienced staff. Just hearing that gave we ex-Massey people quite a lift, and it more than vindicated our move.

Two weeks after joining Ford, Noel Howard called in to see me late one Friday. Checking on how life was going in my new domain and what the people I was working with were like.

"Outstanding," I said.

He went on to say that he had just left an unpleasant meeting at Massey Ferguson, where he had received several veiled threats.

Picking up the phone on my desk the following Monday morning, I was surprised to find it was Noel calling me from Massey's sales office.

Telling me he had just resigned and asking if I could drive across to Sunshine and pick him up. What a pleasure it was for me to grab a shiny new Ford Fairlane, drive to Sunshine into Massey Ferguson's yard and pick up a good friend.

Driving out of Massey's car park I decided to spin the big V8's rear wheels and raise some grit-laden industrial dust. On the way back to Ford Tractors, Noel told me he had accepted a position with Ford Tractor Operations. They had offered him a marketing position and although we would be working with different departments, it was comforting to know we working for the same employer. Over the next few years and as we worked on many different projects, our paths crossed often.

As new Ford Motor Company employees, we spent the first two weeks being introduced to various company policies. Our indoctrination included learning about the current product line up, and we were also given a limited amount of information on a line of soon to be introduced models.

One of the first appointments, was a meeting with Peter Cook from Ford's Industrial Relations department based at Head Office in Broadmeadows. He clearly laid out some basic rules and asked us questions about our expectations.

Before leaving his office, he made eye contact with each one of us and shaking our hands said, "Ford is in business to manufacture cars, trucks and tractors to make the company money. This is done through three ways, one way is to improve our products, so we can sell more units, another is to put the price up, or we can cut overheads. If any of you can find a better way? Please don't hesitate to let us know."

To be working at Ford's small, clean and efficient Coburg tractor plant was a dream compared to the aroma and toxic attitude, I had recently experienced. The difference in attitude had meant my Massey counterparts and I were happy in our work and a spirit of enthusiasm built among the whole group. We were a team.

During 1964, in Australia and also world-wide, we were experiencing a massive surge in Tractor sales. The total Australian tractor market reaching upwards of 24,800 units for the year, had been a number not seen before. Even though Ford had built additional units of the Super Dexta and Super Major at the Dagenham Plant in UK, tractor stocks in Australia early into the new year were running extremely low.

To counter this trend and also to keep the Dealer network viable. Ewen Scott-McKenzie, the General Manager of the newly formed Ford Australia Tractor Operations, arranged for a shipment of the Ford 6000 and Ford 2000 Offset tractors built in North America to be brought in.

Only having seen the 6000 in pictures and other correspondence, I found this tractor quite different to the Super Major. Its modern styling set it apart from anything Ford had released in Australia before, and its size indicated strength. Powered by a six-cylinder diesel engine offering seventy-seven horsepower and coupled to a ten-forward speed Select-o-Speed transmission, meant there would be much for us to learn.

Other features included an accumulator hydraulic system and for the first time in Australia, Ford were offering a tractor with a two-speed power take off. The 6000's transmission was certainly a futuristic move. It was the first shift on the move tractor that I knew of in Australia. We now had an answer to our competitors' seventy to eighty horsepower range and both our sales and service teams were excited, hoping for big things from this radical new model.

While the tractor initially assisted dealer viability, it became dogged with service issues. Finally, for Ford in Australia, customer resistance to its futuristic looks made the 6000 a very disappointing unit. In addition, we had nothing in the range to cover the needs of farmers in the thirty to forty horsepower market.

Introduction of the 2000-Offset did assist our sales in market segments such as row crop farmers and in the market gardens. It too fell well short of market domination. Having only one reverse, and three forward gears, its performance was limited. The lowest gear speed being around three mph was far too fast for many horticultural tasks. The speed range was only one thing that impeded sales. It was pricing, which at first was a little too high for its size that had the most impact. A fact later borne out, when as sales of this unit slowed right down, management dropped the price. Now competitive, sales rebounded dramatically. The initiative meant we cleared all remaining stock very quickly.

USA HERE WE COME

Time was moving along and one day in late August 1964, Jack Walker called John Vaughan and I into his office. As we sat down, he told us that we were going to USA to join a training group who would be training us to cover a new line of tractors. Designated the 6X range these machines were currently being produced in Ford's brand-new Basildon Plant in the United Kingdom.

A wise man going by the name of Murphy once said, "before you do anything – there will be another thing for you to do first."

For us, before we could leave for America, we had to visit the Broadmeadows Vehicle Assembly Plant and see the Company doctor. I was amazed to find Ford had a large medical facility. It had been planned as a fully functioning department complete with a doctor and nurses. They were there and available in case of there was an emergency in the vehicle assembly plant. The centre also had a stock of vaccines for staff that had to travel overseas.

So, it was off to the vehicle plant we went and when there, we lined up to get our jabs. After being pricked we went to leave and as I was reaching for the door, our doctor called out to us.

"You'll probably get a reaction from the smallpox injection in a couple of days. If you do, just stay in bed until you feel better."

As we were returning to the tractor plant, John looked over and said, "my wife's car needs some work. So, I'd better fix it for her to use it while we're away. If I don't come in for work tomorrow, I'll get you to cover for me until I can get back."

John was absent for nearly a week. Wondering where he was, I decided to visit him at home. On arrival, his wife Jo ushered me into his bedroom. His face was a dull green colour and washed out.

When he saw me, he said, "I've been crook as a dog and I've been in bed since coming home after we went up to the car plant."

I never did get to ask about his wife's car, but it took another couple of days before John came back to work.

A few days later, Jack told us to go back up to the Glass House and call into our branch of the Westpac Bank. The bank was located in the basement and following clearly marked signage, we lined up to receive our tickets and travellers-cheques. Unfortunately, before we could get comfortable with the idea flying to the other side of the world, we had to return all of our documents.

We had received word that the Ford Training farm would not be getting any of the new tractors until the beginning of November. So, it would be the middle of the month before they had everything all set for us to go.

Just prior to leaving, the company doctor asked me to come over to the Broadmeadows plant for another injection. However, he didn't have one for John due to his earlier reaction.

While rubbing some alcohol onto my arm before the injection, the doctor said, "You'll need a booster in seven days, so don't forget." After some small talk, he reminded me again as I left his office.

The two of us flew up to Sydney to board a Pan Am Boeing 707 for our trip to USA. We were gob-smacked to find we'd be flying first class; Ford policy at that time ruled that for flights taking longer than ten hours, personnel were to go first class. On top of the flying time were another two days, which Ford determined were needed to allow their international employees to arrive in good shape.

As the sound of those four jet engines came into my ears, I heard a distant voice saying, "I will see that you never get another job in the agricultural industry."

To top it all off, during the flight we enjoyed free alcohol, plus some of the best airline food I have ever tasted. While dining at thirty-thousand feet, I thought about the food Derek Johnson would be eating that evening, but didn't ponder on it for long.

Arriving late in Hawaii we had some shut eye, then down to the beach for some relax and recuperation, prior to gathering up our baggage heading to the airport. Our next leg of the trip would take us to Los Angeles where we would have a short layover and then fly on to Dallas.

After we had been airborne for about an hour, the captain came out. He started walking around speaking to many of the passengers. I was surprised when he stopped at our seats and asked about our journey. While we talked, he told John and I, that this plane was longer than the first ever flight by the Wright Bros.

Sometime later, I remember checking his claim. At 144' 6" the Boeing 707 we were flying in was 24' 6" longer than the distance the Wrights achieved on December 17th 1903.

BIGGER THAN THE BEATLES

Landing at Dallas and as there were no air bridges in those days, we were walking down the stairs. Ahead of us I could see a huge crowd of young people had gathered. All of them were waving, shouting and cheering.

I turned to John and asked, “did you tell them we were coming?”

He shot me a grin, laughed and just kept walking. Reaching the terminal, we spotted a huge banner declaring, “Welcome to the Beatles.” So, the crowd wasn’t there for us. Even if this legion of Beatles fans weren’t there to welcome us, we took it all in good faith, revelling in their applause and hamming it up all the same.

We were checking into our central business district hotel when I remembered the Ford Australia doctor’s words.

“Get a booster in seven days.”

The hotel concierge advised me to walk one block down to the central police station and ask to see the doctor. Remembering his instructions, I broke into a brisk walk and followed the signs down a flight of stairs into a basement full of police. I had to pass through groups of prisoners all yelling their hearts out. A big policeman pointed me to the doctor's waiting room, I entered, found a chair and waited my turn.

After getting the jab, I started gazing around. The place looked somewhat familiar and believing these rooms were similar to something I'd seen on television; my curiosity began getting the better of me.

So, rolling down my sleeve, I asked the doctor, "is this was the place where Lee Harvey Oswald was shot?"

"Yes," he said waving his hand and ushering me out, "just outside my door."

Walking back to the hotel I kept thinking about the whole JFK assassination, and the random number of coincidences that had led me to the Dallas police station.

Next morning John and I ventured out to discover the often over described locations of Kennedy's death. We stopped outside the Texas Book Depository and looking up at the window we agreed it was a long shot indeed. Although visitors were not allowed to go up the six floors and view the spot where Oswald stood, the place had an uneasy presence. I found that taking in all the sombre history was overwhelming and during the walk back to our hotel I felt quite disturbed by the whole incident.

Later we climbed the steps into yet another aircraft. Today, this one was taking us on to Atlanta.

FORD'S FARM

We landed at the Airport of Atlanta, where we transferred to a twin engine, propeller driven aircraft for the flight down to Macron. There we were picked up by Lou Turf. It was the first time we had met anyone in Ford Tractors from overseas and it was Lou's easy manner that relaxed our expectations in what was to come during our first week at the farm.

We soon learned that Lou was a very skilled technician, and that later he would be coming to Australia to assist us by conducting training for the line of new haymaking equipment.

He drove us to a tidy Motel in Cordele where we would stay for two weeks. The Motel was full and at least thirty Ford personnel from around the World were billeted there. After shaking so many hands and after many pleasantries exchanged, it was time for dinner.

The Company farm was in Vienna, Georgia. Quite large and from memory, I believe it was at least a couple of hundred acres. There were a number of buildings housing a wide range of Ford tractor models, many of which were of USA origin and not known to most of the visitors. The greatest part of my interest was the four new 6X tractors located in one of the main buildings. My mind had also been captured by the number of competitive tractors on site, these were constantly used by permanent farm staff for training USA Dealers.

One of the trainers was Wally Burgess, a colourful character who usually had a big cigar in his mouth. He was the guru on industrial tractor-loader-backhoes. Once a week he would grab a TLB of any make and drive up to a hill on the farm to dig a hole.

When I asked him why? He said, “I like to keep my skills sharp.”

In idle conversation after training one day, I asked Wally where he was born in USA, to which he replied, “half way between Oobligoochi, Tulifini, and Weedowie.”

They are names I’ve never forgotten and I smile every time Wally visits my memory.

The new tractor product names at that time were; Dexta 2000, Super Dexta 3000, Major 4000, with Super Major 5000. later these names were reduced to only show the figures on the sides of the bonnet, or as our American trainers called it, the engine hood. Over all it was a smart marketing move, which resulted in less confusion.

In later years I was happy to obtain the services of Wally and bring him to Australia. Once here, we could tap into his wealth of knowledge to run a few tractor loader backhoe training schools Down-Under.

Breakfast and lunch were served daily in the spacious kitchen and dining area, whilst dinner was provided by the motel, or we could visit one of many local restaurants. One of our first visits to a restaurant caused some confusion with one of the ladies who were waitressing. As we were working in the Deep South, I noticed a very actuated twang to the voices of the people in the southern states. A large table had been set up enabling us to sit together. Before long, waitresses, pencils and pads in hand, began taking our orders.

One of them had taken a couple of our orders before she stopped and in a loud voice called out, "All you guys talk funny, and you're kinda hard to understand."

Looking around the room and analysing our group, I could see we had people from all over the world for company. French Canadians, South Africans, Mexicans, Asians, Englishmen, a New Zealander, and Australians, just to name a few in our cluster. They took our orders and after dinner, came back to chat with us. They seemed eager to soak up the stories we told them. Yarns about our countries and our personal exploits. I believe most of them were stories where the truth may have been loosely handled.

On the ranch we were schooled by eight American trainers who were based there, on one of Ford's two experimental farms. The effort Ford had put into preparing the staff for the challenge of releasing this new range of tractors, showed me that the company was serious about the tractor business. It confirmed in my mind that leaving Massey Ferguson had been the right decision and I had a bright future ahead of me.

For John and I, the trip had been very successful. My hold-all swelled with reams of sales information and pads full of notes. Everything that I had taken during the different training sessions was in there. Along with all of this, we had several manuals to bring home too.

While John and I had flown to the United States, two members of our Australian Service Training group had flown to the United Kingdom and onto the Basildon Assembly Plant in England. There they under took similar training. At home as we debriefed for our bosses, it became plain that we were a much stronger and knowledgeable group because of the different training experiences.

NOW THE WORK BEGINS

As both groups returned to Australia and had now reunited, arrangements were made to send four of the new tractor models to Managing Director, Ewan Scott-McKenzie's farm at Tongala. For the next four weeks our team of six, spent many hours going through all of the information gained from our trips. We began planning, rehearsing and perfecting our training presentations. While we were practicing our parts for the service schooling, Jack Walker had been designing his arrangements for a series of one-week schools in major Australian sites.

To get to Tongala, the six of us would travel to the farm on Monday, staying at a hotel in Moama. We spent a month working with the new tractors, sharpening our skills and increasing our knowledge. Getting to and from the farm, we used two Falcon wagons. Les Ward, Roy Pinney, and Peter Newman shared one vehicle, while Colin Knox, Ron Somerville and myself would travel in the second wagon. One Friday, our group left to drive home at the usual time. However, the guys in the other car elected to leave a little later.

Just after arriving home, Jack Walker rang me to ascertain who was in the car and the time I got back. He told me that the boys in the second car had an accident just after passing through Rochester. Luckily, no one was badly injured, but having rolled over to the other side of the road, the wagon was a write off.

On the following Monday, Les and I drove to Tongala to complete preparations in order for us to begin training the first batch of Ford dealer service personnel. The others were broken into two more teams and were also travelling, they were going to regions in other States.

With three teams of two trainers, we began our service schools in early 1965 and instructing with Les Ward, we educated an enthusiastic group of Central Victorian Ford dealer service technicians. Every one of them knowledgeable about the features and benefits of these 6X Fords. Although our primary focus was on service, I could see how a technician's competence could support the sales team. Les and I would feel a sense of relief and pride as we waved off each group of mechanics with a full knowledge of the 6X's proper service procedures.

Each programme lasted five days and by the end of that first training session, I felt satisfied that we had passed on our knowledge. Looking back to those early days, I'm convinced this initiative set the dealer body up for its success with our new tractor range. By the end of our two weeks in Tongala we left with well hone training skills and a sense of accomplishment.

All we had to do now was repeat the process across the continent.

At each venue and before the dealer staff arrived, we would agree who would begin the first two-hour session. It would be covering every aspect of the new engines until we broke for morning tea. After which, the other would spend an hour reviewing the segment just taken, asking questions and providing answers to questions raised by the group.

My session would be on transmissions and go for two hours, with Les taking over for a further two-hour period. We knew from information given to us by our overseas trainers that the training would become monotonous and to keep our minds clear and intensity sharp, we should swap topics as the sessions moved on. It worked well.

During one of the first of Les's sessions he slowly started dropping to the ground, calling out to me in a faint voice,

"Help me Rourkie, I feel really crook,"

He fell to the floor clutching his ribs.

I asked one of the attendees to help Les into our car and drive him to the Echuca Hospital.

After taking all of the training on the first day, our group drove back to the Moama Hotel where we were staying. Les was in the bar having a beer waiting for us, after he had returned from the Hospital. He was acting like a spring chicken, telling us he been diagnosed with three broken ribs. Something that had gone undiscovered by doctors on the previous Friday. My stint at Tongala with Les covered two weeks. From there, it was up to Sydney for the next two weeks and start training with Ron Somerville. Our training circus was now in full swing.

One of the difficulties that had been with me for a long time, was the gut wrenching and nervous feelings I had whenever called to face a group and speak. During that first week in Sydney, the Ford Assembly Plant manager attended my session covering the operation of our Select-O-Speed transmission, a product feature that I really liked explaining.

At the end of the day, he came up and told me, "That is the best explanation I have ever witnessed on any new product."

I confessed my nervous difficulty to him.

He replied, "Ian you must never forget that you know more about this product than any guy in the room, always remember that."

His words struck a chord with me and from that day on, I never felt nervous speaking to an assembly again.

From Sydney, the training schedule took me to Brisbane, where I worked with Roy Pinney for the next two weeks.

Early in the planning for the 6X range release to the Dealer network, someone in the team said we would be needing something different that could not only attract, but draw customers to our Field Day sites. Our brains-trust went into action and finalized on making a triple-coloured barber's pole that we could couple to the PTO of an Select-O-Speed tractor.

On our field day stands, we would disconnect the rear axle drive shaft and after setting up a safety fence around the brightly coloured pole, our demonstration could go ahead. With the engine running, an operator would slowly move the transmission through its ten gears. Each change in the engine noise indicated the shifting of gears. It was an easy way to demonstrate the travel speed and at the same time the tri-coloured pole spun at an even speed. Our demonstrations became quite an attraction, drawing numerous onlookers, convincing many of them to the value of our Select-o-Speed transmission.

After some time, the interest in the barber's pole dropped off. One day back in the office, Jack Walker asked me if I had an idea that we could explore and create a new attraction, one that would draw even more people in than the barber's pole had. Within a couple of days, I suggested that we should build a full-size polystyrene copy of the Ford 5000 tractor. I intended that by using polystyrene we could attach the model to a steel pole and raise it high it into the air. I wanted it to be seen from anywhere on any field day site.

At considerable cost "Polly" was born.

When we first showed her off, Polly was a great sight and we had many people asking us how we were able to get such a heavy unit up onto such a skinny pole. Polly was a big hit and served us well at many rural shows and field days. However, she fell to her doom at Horsham. Horsham is one of the opening field days for the year, and at the time, quite a big deal for cereal farmers across the wheat growing districts in Southern Australia. For Polly though her big chance would be missed. Mistakes were made during the erection process and she crashed to earth, small painted balls of polystyrene filling every crevice of our field day site.

Polly may have died, but in my memory and while I can still hold onto my grainy clipping from the Weekly Times, she will always live on

That same year, I had been away from the office for a considerable time conducting a number of training sessions on haymaking equipment at different Dealer locations. When I returned to the Broadmeadows Plant, I noticed that Jack Walker was not in his office. I asked after him and was told he had joined a Company in Melbourne taking care of their truck fleet.

Jack was a likeable man; who frequently spoke of Mukinbudin in Western Australia. I remember, he could tell anyone who would listen some of the most incredible stories. Then without blinking an eye, he expected all the spectators were believers.

A PROMOTION TO ZONE MANAGER

Time was moving on and by mid-1965, the training had slowed down. New tractors began arriving on dealer's floors and the company's staffing needs also began to change. Being offered a new role and accepting it, found me transferred to a new position of Zone Manager.

The role involved calling on Dealers, reviewing their problems and performance on a regular basis. My initial area was a zone that stretched from Melbourne to Albury and across the Great Dividing Range to Bairnsdale. My territory was made larger by including the whole of Tasmania.

To manage the zone, it was divided into three separate areas, setting aside one week to get around each one. It saw me covering the zone every month. The final week I would spend in our Melbourne office. Here along with fellow zone managers, we would engage with our senior personnel, distilling each dealer's success or difficulties onto an action plan.

During these meetings we would have an open forum to discuss the best ways we could assist each dealer manage their problems and expose their opportunities. Some of the plans were way off beam, but in every case, we tried to put the company and the dealerships first.

I often found myself forming an opinion, that at least one of these senior people could have assisted Ford more by finding himself a new career.

At the end of the meeting, and away from the inner sanctum, one of the other road guys said,

"I once heard a story where the GM of a large dog food company was lecturing his field salesmen on what sales they needed to achieve to up the ante and sell more dog food."

According to my colleague, the GM had told them that to help sales move more product, he had bought the best cans they could find to package the dog food. His marketing team had also designed the best labels, and he had agreed to the best television and radio advertising available. This time there would be no excuses, he expected his sales staff to get out and sell far more.

A snigger rumbled through the sales team.

The GM, beginning to show his anger snapped, "so what's your problem?"

One of the salesmen piped up saying, "The biggest problem is that the dogs don't like the bloody dog food."

It summed up a situation that we could all relate to.

Before my time, Ford had introduced a small scheme of lending money to dealers, who although they were doing a good job, growing their business was being hampered by a lack of funds. One such dealer had been identified by one of our Head Office tractor managers. Without consulting any of the field staff, he had handed over a considerable cash grant.

Needless to say, at one of our monthly review meetings, no one was impressed when we learned that after taking the cash, the dealer concerned immediately went out and bought a farm. Actions like these beggared belief and it was even more surprising that the executive who made such a calamitous decision, stayed on with the company.

In order to introduce fresh thinking, a decision was made to change the Zones for field staff. So, at the next monthly meeting, Colin Knox and I were told to swap zones. We were sitting at the end of the long meeting room table with all the senior Managers crowding the other end. One by one the status of each dealer from my zone was laid out and discussed. This way, Colin could absorb the position of each dealer's situation.

One of my dealers had reached a point where he had a sales penetration of nearly ninety percent, but his second-hand yard was overflowing with unsold tractors and machinery that he had traded in. He still owed the company for two tractors that he had sold and delivered. Despite my previous attempts to point him in the right direction, he had ignored my efforts and tried to continue.

At the time, house rules at Ford insisted that if a Dealer had gone beyond their standard credit terms, the Zone Manager had to hand the brief over to Market Representation to sort it out. Being out of the Zone Manager's control, it was in no uncertain terms that managers were instructed they were not to visit the dealer concerned.

Our meeting progressed with a member of Market Representation explaining how he had visited the dealer and that they had agreed to some incredible plan. A scheme that would get everything back on track again.

He was rambling on, going around in circles for at least twenty minutes when Colin whispered to me, "What the hell is he saying?"

I shook my head.

Colin stood up and in a loud voice said, "Well, is he buggered, or not?"

The managers at the top end of the table were speechless. The man from marketing endeavoured to regain control. However, a further burst of cutting words from Col brought proceedings to a halt and we moved onto the next dealer.

Ford's accounting rules were simple and very effective. All dealer accounts had to be paid by the twenty eighth of each month, otherwise all credit barriers would go into place. No products of any kind, including parts would be shipped. It's my belief that if some of our business's used that proviso today, they would never go bankrupt.

LARGE AND SMALL EVENTS

In February 1969, Noel asked me to assist him with running a school at Merredin in Western Australia. He had gone over a few days before me to arrange for the products to be transported to our dealer's premises. I flew over on the Monday, picked up a company vehicle from our Fremantle plant, then started driving east through Meckering and onto Merredin.

Meckering a small country town in the wheat belt had suffered a substantial earthquake a few months before I passed through. It was a devastating sight.

Looking at the damage, the Council offices' brick walls were straight one side, but had a fifteen-degree downward kink on the other. Yet it was still standing. Many houses had damage, but what did amaze me, was that the earth on each side of the town had been thrust skyward. Rising up at least two metres in places, even over the small hills and making them into a tiered shape. This feature continued for at least five kilometres each side of Meckering.

The two large pipes carrying town water from Perth to Kalgoorlie were broken. The force rolling some of the sections three to four metres away from their original position.

At Merredin, Noel had set up the tractors on the pad behind the workshop where he was waiting for the tractor tyres to fill with water. While that happened, we decided we could arrange our training room, print the work sheets and have everything ready for when the dealer personnel arrived.

We continued working, checking on the progress of filling the tractors' tyres with water. These machines had come from the Perth factory and therefore, carrying out our predelivery checks was essential. Noel and I wanted everything to be ready before the training started.

Twice a day, a heavily laden iron ore train would roar past. The railway line was about five hundred metres away from the concrete pad we were working on. This massive six-inch-thick slab would tremble with every one of the one hundred wagons passing by.

Far different to the trains that wheezed their way through our little town of Burraboi, when I was a kid. These ore trains were hauled by two engines up front, another in the middle of the train and often there would be another engine at the back pushing. One of the team from the dealership told us each wagon weighed ten tonnes, and each one was loaded with one hundred tonnes of iron ore. I managed a quick calculation and reckoned the engines wagons and load was in the vicinity of 11,000 tonnes. It was hard to imagine the horsepower required to move such a mass and get it travelling at one hundred kilometres per hour. A feat of incredible engineering.

After my time with Noel in Western Australia I returned to the zone. There, Colin Knox and I were teamed together to conduct basic training on tractors and other implements. However, we were to place major emphasis on Ford's new haymaking range. These sessions were conducted at Mornington over five days and we presented to between twenty to twenty-five attendees from several different dealers.

Col was an excellent trainer with good knowledge and a keen sense of humour, which made him popular with each group of men we were training.

We worked in two hour shifts over the course of the day, carefully covering all the products. Then on Thursday evening we would put on a sumptuous dinner for everyone. On Friday, we would set a short exam and present them with a certificate to take home.

We had been travelling together from our Melbourne homes to the training venue for about sixty days. The pressure of being away had us weary of the grind, but we were still excited to see new faces and show them around the servicing of these new machines.

Col didn't swear very often, but he did use some very descriptive language at times. One particular day, Col was taking a new group through the names of each part of a reciprocating mower and using a half metre length of dowelling as a pointer. He put the pointer on an oily spot of the cutter bar, then turned to one of the group, asking,

"What's the name of that?"

The poor guy looked down and replied, "a patch of oil."

It's not possible to print here the words that erupted from Col's mouth, plus the fact that the whole class went into total silence.

Col turned to me and said, "what's wrong with them?"

After a slight delay for Col to compose himself, he continued talking over the noise of the still running cutter bar knife. Using his thin wooden pointer, Col continued presenting until he got the end of the pointer stuck against the reciprocating knife. As a result, this wooden pointing tool started getting chewed up. Pieces of three-inch dowl dropping into a crumbling pile and rolling around the floor.

Without hesitation, Col threw away the remaining stumpy end, and turning to me, commanded, “Rourkie, pass me a new pointer.”

Col lost many dowel pointers over the years.

In the early seventies I was sent to New Zealand to conduct schools on Ford’s haymaking equipment. For me it was a three-week stint in the first year and then I went back for another six weeks in the second year. New Zealand is a beautiful country with so much to see on both islands. However, it is such a pity that there is so much volcanic activity and that it keeps most of the Kiwi population on edge.

On our first training trip, Les Ward was assigned to cover the North Island and I took on the other. While Les caught a flight to Auckland, I was boarding my aircraft around the same time and flying to Christchurch in the south of the country. While our flights were nothing out of the ordinary, our arrivals provided us with a new set of challenges. In separate cities we both faced similar problems not of our making. None of our equipment had arrived at its nominated destinations.

After learning this, I contacted Lionel List, the New Zealand Tractor Manager who summoned me to fly to Wellington. There we could discuss what could be done to repair our situation and to get our training sessions back on schedule.

Our plan relied on me flying to Auckland to join Les. From there we were to go to the wharf and if necessary, haggle with the staff. It was important to find a way of swiftly moving the product onto rail transit and forwarding it to the two different training sites.

We sought out the foreman and invited him to discuss our dilemma at the pub over the road from the wharf. Negotiating with him over a *BIG WARFIES' LUNCH* and plying him with a sufficiency of liquid refreshment, we became his newest best friends. Later we could be found leaving the establishment in the company of a very happy Kiwi carrying a carton of beer under his arm. Smiling and slurring, he pointed towards the wharf. All the hay equipment had been loaded onto two flat-top railway lorries while our guest had been enjoying his lunch.

The job now under control, it was onto our separate training sessions.

TIME TO STEP INTO A MUCH BIGGER ROLE

I had been working at Ford for a bit over five years when my old Massey Ferguson colleague, Noel Howard had been transferred to South Africa. He had been promoted to a senior marketing position. About the same time, a new General Manager had taken over in Australia. Clarence E Ragan, had been a senior manager in Ford Tractors in the United States, and Michigan had moved him into a caretaker position at Ford Australia. He would remain with us until a permanent replacement could be appointed.

In November 1973 and about fifteen months after Noel arrived back in Australia, Clarence announced he would be returning to the United States. He was making way for Noel to take over. At the time Noel became the youngest General Manager in Ford Motor Company's long history.

After lunch one Sunday in February 1974, Noel called and asked me to drive over to his home in Templestowe. A drive of little over five kilometres. Arriving and stepping out of my car, I walked over to Noel as he was putting down his gardening tools.

Brushing his hands down the front of his overalls, he pointed to a few chairs in the shade, suggesting that we get out of the blistering summer sun and enjoy a couple of coldies. I told him I'd say a quick hello to Joan before following him onto their veranda. He came with me and after a quick peck on her cheek, we left her toiling in their new garden.

A few minutes of swapping small talk and Noel stood up. He grabbed my hand, his grip firm. Then he said, "Congratulations, you're our new National Service Manager."

Frankly, I was stuck for words, but keeping me quiet doesn't last for long.

The following Monday Noel released a bulletin announcing the change to all of our dealers and internal staff. While I tried to absorb the announcement, I began moving into my much more spacious office. The National Service Manager position had been vacant for some time. Les Ward, who had filled in as acting National Service Manager had accepted another role within the company and shifted to Queensland.

Murphy's Law was at work again and before I could do anything within my role, there were other things to attend to. My first call was to visit

Ron Henderson in Industrial Relations. His office was within Ford Australia's Glass House, it was there that I had to meet with him to formally receive my promotion and be informed of the changes and responsibilities of my new role.

I would now be responsible for the entire service department, the personnel who would be reporting to me and all of the dealer service staff. It dawned on me that my new responsibility meant covering every Ford and Blueline machine in Australia.

For a farm boy from Wakool, it was mind boggling.

About two months later, Roy Pinney who was employed in the Marketing Manager's position, called me down to his office and said, "we have a surprise for you, Ian. In two weeks, you have to be in Detroit for a Service Managers' Meeting."

Things had certainly changed, and in such a short space of time too.

One of my daily routines was to accept 'phone calls from customers that needed to be handled quickly. I also had to digest and reply to the many letters coming in from owners or their solicitor. Most of the time, they were seeking a resolution to their various questions relating to perceived problems with one or more of the Company's products.

Having a good rapport and being able to quickly get through to the selling or servicing dealer, most matters could be resolved amicably. If I received any difficult correspondence or matters that required a legal response. It involved driving over to the Glass House for a consultation with Noel Woodford, the Company's in house Legal Counsel. Noel Woodford, was a great man to work with and over the next twelve months, he enlarged my think tank considerably.

Often, I would receive visitors to my office. One morning while attending to my correspondence, I looked down the corridor and saw my secretary showing two Victorian Police Officers the way to my door.

They stepped in and before I could stand to greet them, the tallest one thrust out his hand, which I took. To my surprise his grip started turning into a strength test. Before this nonsense had gone too far, I eased my hand out of his.

"G'day cousin," he said, "Peter Tampion."

I hadn't seen him for years and after introducing me to his colleague. I, with a wave of my hand indicated for them to take a seat. Peter had reached the rank of sergeant and was working in the traffic department.

"Must be about that time," he said, "now where do you keep the coffee and bikkies?"

I asked Jenny my secretary to arrange morning tea for the three of us. As she started to leave the office, Wendy Orchard, Noel Howard's secretary knocked on the door frame squeezing past Jenny as she came in.

'From Mr Howard," she said, passing me a folded page from a memo pad, "he wants to know your thoughts on this please?"

I looked at the hand writing, it was Noel's and it read, "what are those two coppers doing in your office?"

Smiling at her I said, "There are no problems with that particular subject, but I will come down and go through it with Noel later."

The reason behind my cousin's visit, was that he and his colleague were patrolling between Coburg and Seymour and he had taken the opportunity to call in while they were driving past. After my guests had left the office, I walked down to Noel and ease his mind, explaining the innocent reason for entertaining two coppers in my office. Once he understood my connection to them, the tension drained out of his face and we started laughing about all of the unfounded possibilities he had imagined while they were in the building.

Over the next few weeks and, whenever he was patrolling, Peter would often drop in for a chat, morning tea and bikkies.

The connection to Peter and through him, Victoria Police opened many opportunities for both parties. At one time, I was able to arrange for a squad of police to visit the Ford Proving Ground. As a result, over the years I entertained several members from Victoria Police as I dealt with requests for help into tractor accidents or introductions to people at Ford's proving ground.

After that first meeting with Peter, Noel never again queried a visit to my office by a member of Victoria Police.

FREEZING IN DETROIT

My role with Ford Tractors continued to evolve. In 1975 I flew to Hawaii again, where, as was still company policy at the time, I had a day for rest and recuperation. The following day I would fly on to Los Angeles where I had enough free time to enable me to visit Disneyland on my layover. All I had to make sure of was getting back to the airport by 11.00 pm, jump onto another plane and fly out to Detroit.

The aircraft was due to leave just after midnight, but my scheduled departure time went past with no crew movement, or public address announcements. Eventually the loudspeaker crackled out a message, informing passengers on my flight that there would be a delay. Pan Am had received a call advising that a bomb had been placed on the plane.

Looking out of the terminal window, I could see that there was much activity going on around the aircraft with engineers swarming over every part of the plane. They even had sniffer dogs excitedly running around it, checking every nook and cranny while doing their job.

Finally, an announcement came over the PA system calling us to board, but they also gave permission for people to wait for another flight if they wished. I boarded, confident that as this plane had been fully searched, it would be as safe as any other on the tarmac.

Landing at Detroit, I was picked up by Ken Toze who was the main Service Manager at the time. We walked out to his car and into the snow which was pelting down, I began shivering and had never felt so cold. Seeing me turning blue, Ken took mc to a huge Department store where I purchased a thick overcoat, an item of clothing I treasured for years.

My lodgings were a motel a few blocks past an invisible city limits line.

Ken could see I was new to Detroit and concerned for my safety said, "If you go out at any time, don't venture into this unwritten border, It's extremely dangerous."

As it turned out, I had arrived earlier than expected, so it gave me time to meet up with some of the main tractor personnel.

On the Thursday Ken told me that tomorrow being Good Friday, he would arrange a car for me to use and suggested I do some exploring. I was also quite surprised when Ken told me that this was the first time that the Company would be shutting for Easter. So, on Good Friday morning and in a flash new Ford two-seater car, I drove up to Niagara Falls in Canada.

It was such a cold time; I'd never experienced anything like it. Everything was frozen, ice and snow everywhere. Trees below the falls shimmered in their covering of ice. It felt like an incredible adventure, and I was determined to make every minute count. I purchased a ticket for a walk behind the falls and after donning a raincoat and gum boots, I started following a line of tourists. I supposed we were all intent on the same experience. What I remember from that day is the most incredible roar of the falls. Water crashing down in front of us like some huge curtain of rainbow-tinged motion.

Back in Detroit on Tuesday morning, it was time for me to meet the other service managers from outpost all over the world. It was the first time we had to come together as a team. There were only about five personnel from the USA office present, with another five from countries outside North America. As the meeting progressed through that first morning, it soon became clear that our new World Service Manager was not up to speed with the many subjects that we were debating. However, it was a good meeting and one where were exploring a wide range of issues. Often the topic may have been particular to only one region.

After four days of discussion, it was obvious that most of our objectives had been settled. I believe our Detroit catch ups were important and informative. They were a way of enabling all participants to take away some good ideas and return home with solutions to a few problems.

As planned, I left Detroit on the Friday evening, flying to Las Vegas, America's gambling capital. Although I'd planned a small trip to the tables, my main aim was to fly down the length of the Grand Canyon in a small plane.

To say the trip was fantastic would be a major understatement. My pilot described every rock outcrop on the way out. While he ran through every little change in the detail of sedimentary strata, his enthusiasm for this amazing rift in the North American continent became contagious. Telling me the canyon's dimensions as being two hundred and seventy miles long and in places, it was anything between nine and eighteen miles wide. At times the sides of the canyon could be up to a mile deep.

During the flight, I saw a column of natural rock standing upright among a ridge of smaller pinnacles. Completely vertical it stood around twenty to thirty metres above the base and from a distance appeared completely round and at least seven metres in diameter. Drawing our attention to it, the pilot explained that it was called Castleton Tower and that we were still flying over Utah.

He told me that in 1964 it had been used by General Motors as part of a new model release for the Chevrolet Impala. GM had produced a video of their sporty new convertible, hood down and standing on top of what appeared to be a random rocky outcrop. However, following the long flowing lines of the car, the camera drew away. As it did, it revealed to the world, that the film crew had perched this spectacularly large car on top of this magnificent natural rock column. At the time it had to have been a remarkable sight.

Knowing that GM had used a helicopter to perch not only the car, but a beautiful model in a long-flowing white evening-dress on top of it, must have made for spectacular television. To me, as someone who was more than just a bit interested in the psychology of advertising, this was inspired and very clever marketing.

Also on offer, was a trip for people to ride horses down into the canyon until they reached the Colorado River. It would mean taking a day to descend, then I would need another day for returning. I learned from the brochure that you had to have a doctor's certificate proving your good health. So, with my cup of adventure flowing over, on Sunday I left to return to our own great country.

Returning to my Broadmeadows office, my first priority was a meeting with Noel Howard. I spent an hour or so, briefing him on the people I'd met and the outcomes of our discussion subjects. He smiled and congratulated my initiative after I told him that I had proposed that we should make this World Service Managers Meeting an annual event. My recommendation having succeeded, he seemed heartened when I told him that every Service Manager who attended agreed in setting a timetable for our goals and outcome desires. This would be a tool we would use to measure our progress and successes the following year.

My suggestion having been implemented, the following year our numbers had swollen, and twenty-seven of us were to meet in Detroit for a second meeting. In my humble opinion, this became one of the best decisions ever made by Detroit and one that resulted in major improvements to service visions for the future prototyping of tractors.

A COACH LOAD OF COPPERS

Back home in Melbourne's East, I was introduced to a new member of our Box Hill North School Council. Peter Anderson, was a Victorian Police Superintendent and he asked me if I could arrange for his Accident Appreciation Squad to visit the Ford Proving Ground? This facility is located in the You Yang Ranges south west of Melbourne? The people under his command investigated deaths caused in motor vehicles accidents. I laughed and told Peter that it would be easier for me to get him into Fort Knox. However, he pushed his point, convincing me to at least try.

I had visited the Ford Australia Proving Grounds on three previous occasions and had witnessed crashes of prototype Falcons. Every test designed in order to gain Government approval and allow the company to manufacture the vehicle.

I phoned the manager of the Grounds who briefly told me, "Ian you know the rules, but you can try asking the Chief Engineer in the Geelong Plant."

To make my request official, I dictated an internal memo and asked my secretary Jenny Moloney to type it out. I stressed that we would benefit as a company and the advantages that a visit of this kind would achieve.

Two days later I received a phone call from Ron, the manager at the proving ground.

He said, "Ian, I don't know how you managed it, or what you have done, but my boss has set up a plan to invite all similar Police Squads to attend the next Falcon Test in a few weeks' time."

To say I felt elated in achieving this action, would be underplaying it.

After my trip to Detroit, I had gained a better understanding of gathering data and setting clear and defined goals. This was an effective and structured style of management that I had gained an understanding of. I now knew how important setting data collection streams were and how to drill into the important information they released. Reading my copy of the memo again I was pleasantly surprised at its structure. I figured that somehow my subconscious must have used some of that knowledge I gained in the US, had found its way into the text of my memo.

At eight o'clock on the appointed day, I picked Peter up from his home and we started driving across Melbourne. We were heading south along the Princes Highway toward Geelong. With the southern city showing a little more than a faint outline of buildings on the horizon, we turned west for a few kilometres and from there we took the road up to the You Yang Ranges.

Long before his troops were due to arrive, I wanted to introduce Peter to the wide range of roads and buildings that made up this secretive facility. So, for over two hours we drove around the Ford Proving Ground. I could feel myself filling with pride, as I started showing him each of the testing trials that all of our new vehicles were subjected to. The small boy from Burraboi inside me was amazed at how far I had come, since leaving our little spot on the map.

There was so much to see. Vehicles known to the testing crew as mules are used for trying new components, and several were thrashing around the top lane of the three-mile, concrete fast track. With high banked ends, it had been designed to ensure that a car travelling at high speed would hold its lane, even if the driver removed their hands from the steering wheel.

The company's engineers had insisted on duplicating some of Australia's worst road sections to test their cars on. The facility had a water trench for leak tests, and close to the buildings was a large round concrete skid pad, used for testing tyres and rims. In all, about five miles of dirt road of every surface one would find in Australia had been included. All helping our engineers understand what broke, what didn't and why.

A new building contained a climatic test room. Here they would run a car for hours on end testing its heating and cooling componentry. A concrete pad with round protruding rocks tested tyres, rims, suspension components and much more.

Probably the most memorable test for me, was watching a new Falcon in the compact insulated test chamber. The vehicle's rear wheels were sitting on two steel rollers and as a safety measure a heavy chain was attached to the rear of the car, tethering it in place. A fan was set in front of the car and was ducting forced air through the radiator grille. It had been set to run at the equivalent of 100 kilometres per hour. The Falcon had been calibrated to run on the rollers at the same set speed. Its air conditioning positioned to operate at full blast.

A large array of heat lamps filled the ceiling above the car and had been set to a temperature higher than that ever recorded on the Nullabour Plain. To make this test even more difficult for the car, the engineers had decided to rig the system to duplicate the chamber's humidity to that of the worst ever logged in Tully, Queensland.

A wall of glass separated the car from an engineer. All during the test he would be sitting, watching and making sure all of the data was being recorded onto the computer.

On a previous visit, I had been invited to open the padded door of this chamber, run through it, and climb into the vehicle. The few seconds I took to settle into the roaring car's driver's seat, I found myself pouring with sweat, yet after about ten seconds of sitting in the car's air-conditioned cabin, I felt very cool. Becoming amazed at the efficiency of the car's ventilation system, I started thinking of ways for our engineers to gain access to the facility, if ever the often-mooted Australian built tractor cabin came to fruition.

Just after lunch on our appointed day, a police bus filled with a happy group of coppers arrived in time to witness a crash test of the new vehicle. Earlier, this hand-built Falcon had been loaded with a number of crash test dummies and had been positioned about twenty metres from a substantial concrete block. The face of it flat and the rear angling down at forty-five degrees, I thought it would take a good bulldozer to shift it, that was until I found out its foundations were even more huge.

A cable from a petrol driven winch mounted behind it, ran through a hole in the base of the block and was attached to the front of the doomed Falcon sitting on the tarmac approach. In front of the block was a pit similar to those you could find in a motor mechanic's workshop. On either side of this pit, high speed cameras were mounted to record the crash in minute detail. Triggered to provide a photographic record of each stage of the impact, these images were vital in enabling Ford's engineers to understand the sequence of damage.

One of Ford's white coated test engineers started telling the police group that this trial would involve towing the Falcon at a designated and exact speed until it smashed into the block, stating that if an error arose at the start, they would abort the test immediately. Suggesting everyone concentrate by locking their eyes onto one spot on the car, he explained this would be their best way to witness the panels folding and crumpling, as the car's inertia dissipated into the concrete.

Tension filled the air until after sounding the, "All-Clear," the Falcon began on its first and final journey.

When the sound of smashing glass and tearing metal subsided. Ford's engineer explained how the predetermined design for the crumpling effect had worked, saying how pleased he was that certain components had broken out at preset angles. Pointing to the engine, he explained how it had moved down and rearwards as a load of fierce deceleration forces happened. The tail-shaft had broken away from the differential as planned and the damage continued. The crash test dummy's chest had collided with the steering wheel, the force twisting and compressing it. The front fenders had, as their design intended, crushed and splayed out at an angle. Even small items like the dash mounted switch knobs were planned to break off.

Spectating with us were several government engineers. They were present to witness the test and verify the results, all needed to approve the vehicle for production. Ford's engineers went into much detail as they explained the way the Falcon collapsed through the inertia and how over the next months, they would examine the photos to make sure their predictions were correct. If weaknesses were found, these would be areas to redesign and improve. For both Ford and the Victorian police force this would become a working relationship to build on.

After we bade goodbye to his colleagues, I took Peter on an extensive tour of the facility.

For me the proving ground had always been amazing. Talking to Peter while I returned him to his home, I could see he found value in the rigid testing that the company had invested into the building of their cars. A few days later, he told me that his colleagues had also found the exercise extremely worthwhile, and one which gave the traffic branch a better insight into the way a car reacts during a collision.

Following the visit to the proving ground, I occasionally had visits from Victorian Police officers from the Accident Appreciation Squad. They were keen to ask me for my thoughts involving aspects of different tractor accidents.

One accident of particular interest to me, involved a Ford tractor fitted with a front-end-loader. The machine had overturned while traversing the side of a hill. I learned that the driver had filled the bucket and without lowering it sufficiently, had been reversing downhill when the tractor rolled over, killing him.

It was certainly a tragedy and I felt for his family, but this was an accident that may have been avoided. An assumption the officers agreed with.

I asked them to provide the details from the tractor's identification plate. With a quick check of our records, I discovered that this machine retained its original engine, but had been fitted with another transmission and a different rear axle.

For me and the officers sitting across the desk, it meant we were able to close the case.

TSBs, WARRANTY and the RPM MANUAL

The Australian National Service Manager Tractors role was a formidable task and it took time to adjust. My responsibility now covered the needs of every of our one hundred and seventy-five selling and servicing Ford Tractor dealers. Luckily, I had two fantastic assistants, in Jim Dailey and Kevin Gascoigne.

On a serious note, mine was a senior position in a huge Company. A multinational business, which had almost eight decades of expertise that I could call on. From my first Service Managers' meeting in Detroit, I knew should ever a new concern arise, I could call anyone within the company. Colleagues from whom problem solving knowledge would always be freely given.

Under normal circumstances, all service directives came from the United States. However, on a local scale, assistance was always at hand from our Automotive Division. Also available to Ford Tractors was an internal telephone system that linked all of the Australian Ford Divisions together. So, if anyone had a question on a product and it needed an answer from outside the office, it was only a question of using the phone. It certainly beat going over to the vehicle plant and speaking to someone that had that special knowledge.

In some cases, I needed to visit the metallurgists in our Geelong Plant to get an answer for the failure of a variety of metallic parts. Another area of valuable information came from our Australian suppliers. Companies like Dunlop, Olympic, Repco, Castrol, Atlantic and BP, are just a few. Experts from each of these companies welcomed my visits, and they helped our team find answers to the cause of many unusual failures. Although I have always been confident in my own product knowledge and mechanical ability, it was sometimes prudent to search out another opinion, which would either confirm, or deny my findings.

In addition to conducting Service Training in Australia, we issued official Service Bulletins to our dealers. Known through the Ford community as TSBs, these missives contained important information about product updates and improvements that arose from time to time.

We also drafted and issued another publication containing local service information. These Service Memorandums, were written and collated by my service group, pulling all of our Ford and Blueline products together in one place.

Seeing a binder marked Service Memos on a dealer's shelf, or in a technician's service van made me proud of the publication and the great work my team were doing.

Sometime in the seventies and during one of my many visits to the Glass House, I learned that our automotive division was designing a new warranty system. It would be a greatly improved process and where the old scheme needed a large group of people to manually approve each claim, this one would be computer driven. To manage the change, the Company had built a new structure to house a computer large enough to handle Ford's future warranty needs.

As our Tractor Division had entered a new era, our own warranty processor, Eric Thomas, soon became overloaded. It was not due in any way to product reliability, but in a few short years we had moved from two basic tractor models to a wider product offering with two transmission options with dozens of different parts and options. Together with the two main Industrial rigs, the Ford loader backhoes and overseeing Blueline claims, Eric was working longer hours trying to keep ahead of his work load.

With pen and book in hand, I headed to Head Office and located the executive in charge of the entire warranty project. After an extended discussion where I outlined the pros and cons of introducing tractors into his system, he agreed to our inclusion. Telling me the criteria would not be difficult to add, he suggested an eighty-line data system, which would easily cover the proposition I had raised.

At Ford an in-house system was in place whereby each division paid fees to use one of the other groups talents. For example, our tractor department paid the assembly section to build complete tractors from skid units as we required them. It also applied to the National Parts Department which charged our sector to stock and distribute Parts.

Therefore, when the new warranty scheme came into being, Noel asked me,

"How much is this new system going to cost us?"

To say Noel was taken-a-back when I told him we would not be paying for the scheme, but there would be a small fee applied for staff key punching each claim into the system. The charge for this action was very low and a very surprised Noel was more than happy that we'd had a win.

There were a number of coding fields put into the system that were unique to Ford Tractors. This ensured that any claim with errors, or dubious items could be referred back to our service area for further investigation. We were able to insert a code that identified a claim that had been submitted for a tractor which we knew to be stolen. When this occurred, the claim would then be sent back to our office where we would notify the necessary authorities. An extremely effective method for identifying theft, the inclusion of this single code is something that still makes me very proud.

Although our Operator Manuals provided clear operating and maintenance schedules, I believed there was another opportunity where we could gain an advantage in a crowded Australian marketplace. If we provided a simple method to new and subsequent owners of Ford tractors for recording the service history of each machine, it followed that the used value of Ford tractors should improve.

From a company perspective, these proper service records would head off some of the out of time claim arguments we sometimes experienced. We were confident our dealers were competently stating the terms and conditions for warranty at time of the tractor's delivery. However, it's not always something that's completely understood. I wanted our clients to know exactly what was, and was not, covered within our warranty terms and how they related to the Australian laws that applied.

Under an Agreement with Ford, our Operator and Service manuals were stocked and sold by a private company based in Moorabbin, relieving my department from the responsibility of distribution. However, our sector still had responsibility for all of the editorial content. As a result, our members of the Service Division spent many hours writing and editing local versions of operator's manual. Every book had to be checked and double checked before being released by the printer, and ready to send out with each new tractor.

I identified that we had both an opportunity and a need, and it was from this, that the RPM Manual was born. An acronym for, Regular Planned Maintenance, the title may have been simple, but I believed this to be an important tool we could exploit to gain an advantage in the tractor market.

The task was immense and compiling a service schedule for the life of a tractor meant checking the validity of a million variables. The need for a servicing schedule was nothing new and many operators' manuals from every tractor company had a brief version of my idea. Through the eighties, motor racing teams began adopting a similar system for determining the life of service parts, and our RPM manual also had these inclusions.

When these additional manuals began showing up with each new tractor arriving at a dealership, I had been hoping they would be met with optimistic enthusiasm. The dealers had been introduced to the idea at service meetings, but I found myself hesitant to call it a roaring success until much later. My doubts were unfounded when I learned that more than one dealer saw it as an opportunity to gain extra profits through the sale of service work and spare parts.

Many people worked to bring this unique manual to the Ford tractor buying public. Although Ford New Holland's new service team more or less buried the publication, collectors of Ford tractors from the era have made it a much sought-after book. While my team did most of the grunt work, thanks also must go to Noel Woodford who wrote the whole legal section on Warranty.

During my time in service, I often assisted members of the Fordson Tractor Club of Australia Incorporated. Sometimes these may be technical information inquiries or model codes that their members required. I feel great pride every time I think about my interactions with "The Club" and knowing that it has been a huge success story since its inception. I believe this must be credited to the passion and management by a capable Board, who have been ably assisted by Dawn and Barry Milson.

In 2016, I was very proud to be inducted as an Honorary Member into the Fordson Tractor Club of Australia. More thrilling has been joining my old colleague Noel Howard, and South Australian tractor dealer Ian Denton, on the club's honour roll. Friends whose dedication to helping the club and its members, had been similarly recognised.

From left Dennis Greenwood, Ian O'Rourke, Noel Howard, Rob Bridger, Ian Denton, Ken Rice.

A GOOD MEMORY HELPS

From time to time in life, we all experience some quite unusual things, it could be the tone of a voice, the profile of a shadow, or a mention in a conversation and memories come flooding back. One of those moments I clearly remember, happened when I was heading from my office down to the Marketing end of the Company. I passed Ian Lauder, the Ford Finance Manager and a man with his back to me. The stranger's voice began trickling through the filing cabinets of my brain and after a few seconds, I turned back and broke into their conversation. I could see in the stranger's face, that as he started looking at me, he recognised me too.

"Brian Teelow, I'll bet?" I said, "Deniliquin High School, 1948."

Both deep in conversation before I interrupted them, Brian told me he and Ian had been talking about financing a new Ford tractor. So, I shook his hand and left them to it.

Later that day, Brian came down to my office where we had a good chat about the old days. However, we wanted to know more, so we spent time we didn't really have, sharing and discovering all the things each of us had done after our school years had finished.

Another Friday I remember well, is one where I received a 'phone call that came to me from Jack Luhrs, our dealer in Griffith in New South Wales. He had a couple of queries that needed more attention than just a phone call, and convinced of his concerns, I told him I would drive up to see him early the following Monday.

Arriving at the Griffith dealership by 11.00 am, we continued discussing Jack's concerns until around 12.30 pm, when he looked up at the clock above the visitors' chairs and suggested we break for lunch. Deciding to head to a local hotel, I accompanied him to his car. After a short drive, Jack pulled up beside a paling fence in the hotel's car park.

As we arrived in the hotel's car park and before I began stepping out of the car, Jack pointed at the boundary and said, "see those red marks on the fence?" his index finger drawing circles in the air and showing me where to look, continued. "Well, that's the blood from Donald MacKay." There was now a tone of morbid curiosity in his voice. "You know, the anti-drugs campaigner. He was murdered just here!"

MacKay was a Griffith local who owned a furniture shop in town and his agitating had been stirring up a number of people, who it had been alleged, were growing Marijuana on one or more of the hundreds of farms in the region.

MacKay's murder became a big news story for many years and to this date, his body has never been found. However, because of his complete disappearance it has spawned many different and sometimes outrageous theories. For the people of Griffith until his body is found there will never be closure.

During times of travel, most people remember meeting someone who for a lot of different reasons, seemed to stay in their memory forever. For me it happened on a return trip from London to Australia. The Qantas jet touched down in Athens where we were taking on additional passengers. During the first leg of the flight there had been an empty seat beside me. Not taking much interest in the throng of faces all pushing toward their seats while waiting for our new passengers to board, my interest soon changed as a gentleman excused himself, taking the empty seat beside me. This well-known Australian actor at that time was at the height of his fame. As he settled into his seat, we shook hands and I enjoyed a most memorable conversation until touching down in Sydney.

I can still recall a lot of our conversation and listening to his stories. His name was David Gulpilil, and a very, very famous Australian actor. For a boy living in a mud-wall house from Wakool, it sometimes makes my head spin to think about the people I've met, and the places my career has taken me.

On another occasion, I had been catching a Jumbo from Melbourne on another flight taking me to North America. Having made my way through to the boarding lounge and onto the flight, a well-dressed man, tall and with a Texan twang sat down beside me. We soon engaged in conversation and me being a talkative type, introduced myself and asked him for his name. He told me he was J. Robert Erwin, a quick search of my mind didn't ring any bells in my memory banks, but it certainly did when he spoke about his Dallas based company and the products they produced.

He said, "Our company invented Liquid Paper."

That certainly aroused my attention, that product was on every desk, in every school kid's pencil case in the developed world, and many undeveloped ones as well. His voice definitely had that Deep South accent and he had an air about him, that little something extra, which is difficult to describe. A few years before, I had read a comment relating to a person as having a voice like condensed milk flowing over gravel.

That quip covered J Robert Erwin to a tee.

Meeting people on planes is a consequence of travel in Australia and, like Forrest Gump says, life really is like a box of chocolates. You never know who you are going to meet.

On this occasion, I had spent five days visiting dealers in mid New South Wales and the toll of travel was telling on me. I arrived at the Tamworth airport in plenty of time to board a Fokker Friendship aircraft heading to Sydney. From there I would catch another flight to Melbourne.

Being mid-afternoon and the end of the week, it seemed everyone wanted to get home and the plane quickly filled. My seat was on the rear row with an aisle allocation. I settled in and was grateful to find an empty window seat beside me. I remember looking at my watch and thinking we must have been close to take-off. What was the pilot waiting for, the plane had a body in almost every seat?

My answer came in the form of a lady with a very high-pitched voice entering the front door of the aircraft. Taking her time and gradually moving down the aisle, she seemed oblivious to the pressure of travel. While I kept thinking about how long it would take to get home, this television icon was chatting to everyone as she moved along.

From my position it was difficult to see her. It looked like she was holding up half a wattle tree and as she arrived beside me, I stood up and moved into the aisle allowing her to get across to the window seat.

Her greeting was her well known catch-phrase, "hello darling," and she was the outlandish television celebrity, Jeannie Little.

I can still remember one of her television stunts. Out to make an impression and, probably a point that's lost on me now, she entered the studio wearing a dress completely made out of black plastic bags. Jeanie Little was incredible.

All the way to Sydney she had much to say, and for all of the flight, she kept a death grip on her small wattle tree. She certainly was an all-Australian celebrity.

Over the years, I have met dozens of people from all over the world and have made a habit of saving their business cards into a three-ring binder. These cards now number in the hundreds and can chart the career of more than half the people I did business with. They've also proved a great resource, as many of these people from all across the world had been, or still were, involved in many of the businesses that could help me to solve an issue if I needed them.

Staying in touch has been a very successful habit and has often proved useful on one or more difficult occasions. One such incident happened in the late seventies, a number of Ford Tractor Dealers began advising our Service Office that the oil enclosed disc brakes on various tractor models were experiencing, "brake chatter." This phenomenon emerged as a deafening noise emitting from the tractor's brakes when applied. Our original internal investigations did not bring forward any answers and none of our local oil companies could find a solution either.

Beginning any investigation, the first clue is to ask what has changed? Checking with our engineering departments we were able to find that the materials used to manufacture brakes on all of the models involved had not changed. Therefore, it had to be an alteration of compound within the oil, which would be difficult to trace, especially when our oils were coming from several, different and reputable companies.

One of the oil companies came up with a possible clue, a chemical that had once been used as an additive in oils had come from whales. However, with a shift in the public's acceptance around the harvesting of these majestic creatures and legislation banning the trade in any products derived from whale slaughter, they reasoned this to being a hint to our dilemma. Incoming correspondence proved that most of the oil companies had tried to resolve the brake chatter problem with different additives, but reported that they had experienced only limited success.

I turned to my cache of business cards and found the name of a highly regarded additive company, Lubrizol... They had an office in Melbourne so it was quite easy for me to contact their switchboard and get through to the name on the card. After laying out the problem and some replies from others seeking to assist me, one of the chemists at Lubrizol suggested one of their products that he thought might help me. A small sample of the new Lubrizol additive was supplied to Ford, this we quickly sent to two Dealer's to trial. Success was immediate and, at the same time, we were able to get our oil supply companies involved.

Another habit I developed, was collecting and exchanging cuff links and tie bars with my

Australian colleagues along with many other company and random people I met during my travels. Most of them have come from personnel in different departments of the Ford Motor Company, but some more collectable ones have come from the individuals who made them. Others are from businesses that are no longer trading and as well as being keepsakes, these are very sought-after items. Keepsakes that are becoming more valuable as time passes.

HI-HO, IT'S OFF TO FRANCE WE GO

Around the middle of 1975, my colleague from the industrial side of the company, Vin Smith and I were called into Noel's office, Noel met us with the news that Ford had bought Richier, a French Company manufacturing excavating machinery, tower cranes and the new Ford wheel loaders. He sent us to the Glass House where we were to arrange tickets, travellers-cheques and documentation for us to fly to France. There we would be trained and provided with enough sales and service information to conduct product schools prior to introducing these machines into Australia.

Our trip from Tullamarine was an easy flight and memorable for nothing more than the weariness that comes from an international flight. We landed in Hong Kong where we transferred to a hotel for an overnight stay, giving us almost a full day before boarding the plane for an overnight trip to Rome. Overjoyed that our itinerary also allowed us a day in the Italian capital for a day's rest and recuperation, we made sure we arrived back at the airport in good time to catch the Rome flight.

At the airport we waited while the British Airways plane that we were to travel on landed. At 5.00 pm and as it touched down, a flap from one of the wings fell off, grinding, sparking and sliding its way down the tarmac. British Airways had a problem, this bird wasn't flying anywhere until it had been repaired.

For Vin and I, it meant a taxi ride back to the hotel and spending another night in Hong Kong. We were delighted to have another opportunity to explore more of Hong Kong's culinary delights before getting back to our rooms and booking a wake-up call. We would lose some time on the overall journey, but we still had plenty of time to check in and board. Our new flight was scheduled to take off at seven o'clock next day. So, it was Hello Hong Kong tonight and Rome here we come tomorrow.

Before leaving our Rome hotel the next morning, we had booked a cab to chauffer us to around the city. We couldn't believe the irony when our driver told us his name was Lucky Luciano. The American Lucky Luciano had made a name for himself as a gangster running rum and some say the original Mafia Godfather.

Our Lucky knew the connection and laughed about it. We took an instant liking to him, he had a great sense of humour and, as we drove around, made it his duty to keep his two crazy Aussie passengers informed about the places and people from Rome's rich history. We ended up visiting all the tourist traps like the Vatican, Lucky describing the Pope's residence as if he were a regular at dinner on Sunday. Taking us to see the Sistine Chapel he had us wishing we had binoculars, which would have afforded us a better view of this famous ceiling painted by Michael Angelo. Although we kept teasing Lucky by insisting it had been an Aussie painter and decorator, Leonardo Da Vinci from Broadmeadows who did the brush work. Lucky ensured we had a great time driving around the Colosseum, even convincing us to toss three coins into the swirling water of the Trevi Fountain.

That night it was off to Paris on another jetliner, then onto yet another plane before reaching our destination in Charleville. Once there, we were to gain information on the excavators and tower cranes and build a report for Noel. What we found amazed us and our first impressions were far from the images we had painted in our minds. The factory must have been a target during World War II, because a couple of walls had huge shell holes in them and even though the war finished twenty years before, it hadn't been a priority to repair the factory. I guess they were lucky to have a roof.

The hardest time we had, was putting up with the instructors smoking dark tobacco cigarettes, plus their need to drink Beaujolais over lunch. Not good in my book when we were trying to learn about new products. On the Wednesday afternoon we boarded a bus to take us to Lyon, with a two hour stop at Reims. There we had been invited to learn a little more about champagne. After sampling our share of the different wine being offered, and filled with more than a little happiness induced by the tastings, we reluctantly climbed aboard the coach.

After Reims it was another bus trip to Lyon where we would spend two days learning about the new range of wheel loaders with Ford representatives from around the world. These new Fords were designated A62, A64 and A66. We didn't know it then, but these units were to have a big impact on the Australian market after released for sale.

At the end of the week, we joined our colleagues in a touring coach for driving back to Paris. Once there, our itinerary allowed enough time to tour the capital sight-seeing. Vin and I stood in line and bought tickets so we could go up the Eiffel Tower and take in the sight of Paris from above.

Unfortunately, and even though they happily took our money the fog had started rolling in. Deciding that we were weren't going to see much from the top, we only took the lift to the first floor. That and exploring Paris a bit more took care of Saturday and our French sojourn was quickly coming to an end. On Sunday we hailed a French taxi which took us to the airport and off to London for another week, where we took in product schools on the new machines. By the end of the training in Britain, we had become tired of the travel and looked forward to flying home.

An airliner's seat was never so welcome.

Arriving home, I found Noel had scheduled a meeting to explore the opportunities and pitfalls if we were to introduce tower cranes and excavators in Australia. Following a vibrant and willing discussion, we agreed that the cranes could not be sold here. What made it unsuitable was the main frame. Being horizontal, it would infringe into any neighbouring air space and therefore machines of the Richier design were not permitted to work in Australia. While disappointing in some ways, cranes would take the tractor company into a completely different market. A market where unknown risk could have had a detrimental effect on the company's overall profitability. Therefore, we decided to stick to what we knew, tractor-based machinery.

Incidentally, although we as a company passed on the product, seven Richier tower cranes had been used in the construction of the Sydney Opera House. On completion all of the cranes were dismantled and shipped to Perth, I lost track of them and am unable to recall where they finished up

We explored the costings and opportunities for the excavator products and after a lot of work from the people in finance and sales departments including spare-parts, the risk outweighed the reward. Even though the excavator market was beginning to gain traction in the building industry, as a company we came to a decision that distributing excavators would not be viable. Although the opportunity to make a fast dollar may have been there, to proceed with that product line, would have meant a larger inventory of spare parts. Coupled with the ongoing costs of servicing the needs of a specialized product in line with Australia's consumer laws, it made sense we passed on bringing them into the country.

We did agree that the opportunity for sales of the wheel loaders was huge. Over the next few weeks, any spare time I had was taken up with planning the service training and a myriad of other elements prior to their release to our Industrial Dealer Group.

Within the loader range we had three models to plan for. These were the Ford A62 which we imported with a one cubic metre bucket. The A64 came complete with a two cubic metre unit. The

A66, we supplied with a three cubic metre bucket, giving it a huge material handling capacity for the time.

All of these models were mainly derived from major components of the 8600/9600 line of tractors. As well as being very popular products, the commonality of spare parts meant there was a smaller inventory outlay for both company and dealers. Being of a modern articulated design and having an up-to-date ergonomically planned operator environment, soon saw the heavy loaders selling well. These 'A Series' loaders were now satisfying a long called for need by our existing Industrial Distributors in Australia.

To my knowledge there was only one major problem with a unit in Western Australia. The selling dealer reported that a customer's A66 steering had locked while he was building a dam and the steering wheel could not be moved. As any malfunction of any Ford product involving safety must be investigated immediately, the dealer was advised to bring the loader into his workshop and await further instructions.

One of my colleagues called Engineering in Detroit and spoke to Joe Abrahamchik. An engineering chief, Joe had been responsible for the design of steering and electrics of these loaders. His immediate decision was to fly to Australia and after arriving, I would accompany him to view the machine. I advised the dealer Joe would be arriving in Melbourne in three days, then we would both fly to Western Australia to check out their machine. When we arrived, the loader was up on a hoist to make it easy for us to view the offending part.

After rubbing at a number stamped into the steering box Joe looked at it for a couple of minutes before saying, "this is not the correct assembly."

During our discussions with the dealer's service staff, we established that the day before the failure, the owner had reported an oil leak coming from the machine's steering box. Knowing the importance of keeping the loader in service, they ordered a new part from Ford Parts Depot in Broadmeadows and had fitted it. Although the machine would be out of order until the right steering unit could be found, we knew it was a simple part to install and the machine would soon be back in work.

On returning to Melbourne, we visited the National Parts Depot and checked the four items remaining in stock. Although they all appeared identical, Joe declared them suspect, and selected two out of the four, declaring they were the same as the failed unit in Western Australia. A closer inspection revealed that there were different part numbers marked on the housings. The only explanation we could accept, was that the original manufacturer of the steering boxes had supplied two incorrect parts in the batch they sent to Australia. An immediate action was implemented to withdraw the offending part from all stock held in worldwide Company Stores.

Within the week Joe returned to the USA. However, he was a wonderful ally in the service of Ford and I made a special effort every time I went to Detroit, making time to renew my acquaintance with him.

Looking back over my career at Ford, I am still very disappointed that the company dropped the wheel loaders from sale. I believe we missed a trick as they were still very popular and very competitively priced in their market segments.

THE BEGINNINGOF THE 8401 PROJECT

Sometime in 1978, Noel invited me to join him on a walk through the tractor assembly plant. I was thinking he just wanted to stretch his legs and discuss service issues while he walked. However, he led me to a huge white canvas screen in an unused corner of the factory. Behind the screen were four new tractors. Split at the clutch housings, an 8600 and a 7700 stood side by side and were open between the engine and transmission. I could only guess what he had in mind.

I could see Noel watching me closely, and after giving me a couple of minutes to take it all in said, “you can’t talk to anyone about what you are seeing here.”

I agreed and listened to him say that because Chamberlain had a 100-horsepower model selling very well and, as we didn’t have a model in the pipeline to compete with them, he intended to switch the engine from the 8600 and fit it to the 7700 transaxle. He understood we would need to reduce the power of the 8600 by fifteen horsepower, but a new fuel setting plus changing the injectors would take care of that. Noel emphasised it was important for the project's success that the engineering division, or anyone else in Dearborn, were not to be told of his experiment. This was a Ford Australia project and under no circumstances, could any division outside our select group find out what he was planning.

Even though I thought it would be a great idea to have a 100-horsepower tractor in our line-up, my mind was boggling with other thoughts, not the least, reliability. Ford had engineered a lot of margin into our transmissions and by doing so we had an enviable reputation when compared to our competitors, but putting a six cylinder in front of a four cylinder transmission had me worried.

I told Noel that the 7700 rear axles wouldn't be able to handle the extra power as the three pin epicyclic reductions in the rear axle housings were likely to fail. I reminded him we had already seen a couple of failures where oversize tyres had been fitted and then filled with water, in the field. It might have been a disaster of the customer's design, but it was a failure of the three pin reduction gears all the same.

"Well," he said, "in the interests of the Company we are going to do it." and within a few days the two unnamed test mules were ready to roll. All products need a name or something to distinguish them from other models and in this case, the team came up with 8401.

The name Eight-Four-Oh-One, was derived from the number eight being part of the 8600/9600 over one hundred horsepower group, from where the engine was sourced and 401 being the engine's cubic capacity in inches. It was easy to remember and stood apart from the other models. This tractor would be all Ford and created in Australia

Both test mules were put to work in the field not far from Broadmeadows However, a few days later a troubled Noel came into my office with bad news. He came looking for a solution. One of the 8401's had broken a rear axle housing, I had no complete answer as to a cure, but suggested the test team repair it and take all the water out of the rear tyres. I knew how much he wanted this to work and I didn't want to be a prophet of doom, but in that moment, I didn't have a fix for him. Noel knew that the next day I would be leaving the office to fly to the United Kingdom to attend our yearly World Service Meeting, so wishing me a speedy return, asked me to think about it while I was away.

I had been looking forward to catching up with my colleagues and the new Chief of Service, Bill Hepburn. Bill was a strong and knowledgeable leader, someone very suited to the position. His initial move was to take all of us into the Engineering Design Centre, a place that rarely accepted any personnel other than their design staff. It was an exciting change and one we were all keen to embrace.

Once settled in, we were introduced to John Foxwell, Ford Tractor Operations, Chief Design Engineer and there right in front of us was a prototype Ford's new project. Introduced as the mock up for a new 8210, it was almost a copy of Noel's idea. They had bolted a six-cylinder engine to a 7700 transaxle.

John welcomed the group and began talking about the prototype in front of us, expanding on the company's huge hopes for it. As I knew what was happening back in Australia, I listened carefully to every question being asked and every answer being delivered, this was most intriguing. Although he couldn't give us a launch date, John said the machines would only be available for sale after completing a regime of rigorous testing.

It came as a bit of a shock when John invited us to walk around the 8210 and make suggestions. Such an action was unheard of in Ford Tractors, soon service managers from across the globe swarmed over this shiny new machine. John welcomed our exuberance for making constructive comments. Knowingly he had primed several engineers who stood ready, clipboards and pens on hand to take notes.

I purposely walked over to John Foxwell, determined to discuss one or two of my pet problems. I began by showing him a photo of a 7700, it was working an offset disc on a dry rice field in the Griffith area. All that was visible of the tractor, was half of the front wheels, and red dust engulfed the rest of the unit. Everything was covered in fine red dirt.

I asked him how he thought we might improve the cabin air filtration system. At this time the air conditioning intake was drawing air and dust from just above the tractor's rear window. I described how dust, up to an inch deep, hung on the window glass and because of the forward motion of the tractor, this would always be a low-pressure area.

I also explained that to overcome fine dust choking the air-filter, we had tried improving the problem by fitting a 90mm plumber's pipe from a plenum at the rear, sweeping around the side of the tractor cabin, where it could draw air from just above the windscreen. Although not the perfect solution, it had proved to be a step in the right direction.

John acknowledged that we did have a product concern in Australia and said, "I will talk to you about that on Friday."

A fresh-faced engineer came up to me and said, "Ian, I think I can solve your problem." with his voice full of confidence, he continued. "If the tractor is travelling north to south, tell the owner to drive east to west."

I asked him if he had ever seen, or understood the benefits of contour banks.

His answer was a polite "No!"

Using his notepad to draw diagrams, I explained the features of this moisture saving concept and how Australian farmers worked their tractors between them. As I saw him beginning to comprehend the benefits of this dryland farming technique, I started thinking, maybe that young man learnt something that day.

In the meantime, the rest of our team were looking carefully over the 8210, providing a number of helpful suggestions that were dutifully jotted down by the clipboard carrying engineers.

For me, the most intriguing aspect of the 8210 project, was the cabin windscreen. It had been covered in coloured lines with black dots and it also had red lines in various places. A closer examination of it revealed that the glass screen was embedded with operating data such as the engine revolutions, ground speed, gear pattern, engine oil pressure, radiator temperature, and many others I cannot remember.

One of the team, paraphrasing the famous Professor Julius Sumner Millar asked, "why is this so?"

We were told this was a new testing development allowing the operator to keep his eyes on the front ground area, while at the same time, see the main operating data of the tractor. It was what we came to learn as a heads up display.

Incredible! Then someone asked, "what happens when the tractor is travelling along a main road and a passing vehicle throws up a stone and smashes the windscreen?"

The engineer gave him a quizzical look saying, "we must review that."

My attention now turned to the rear axles on the 8210, so I asked John if they had upgraded the assembly to cope with the increased power being fed into it.

He replied, "we have employed a four-pin planetary for better strength and to increase reliability."

I waited as he explained it and then asked, "will they fit the standard 7700 axle shaft and trumpet housing?"

His reply was a definite, "yes!"

I thanked him and without saying any more, continued to listen to the questions and answers being bounced around the engineers and our group. During our lunch break I went over to see the Parts Manager, Paul Cherry and asked him if he had any of the four pin planetary gears on hand.

He checked his computer and said, "yes, we currently have one hundred sets in stock."

I gave him the fictitious order number, IOR-1 with the notation, "Air freight to Australia." Later that night, I 'phoned Noel with a solution to our rear axle concern. Concern was a term I preferred to use when dealing with dealers and owners as it seemed to take the tension out of normal language. Early in my career, I began calling mechanics technicians and it had become another one of my ways for elevating the image of our service personnel.

Later in the week, our group were invited back to the Design Centre by John Foxwell, who proudly showed off the suggested changes made to the 8210.

To my surprise there were two shiny white pipes going around each side of the cab with a large filter above the windscreen. John turned to me with thumbs up. What a deal, here was for the first time in my experience, an engineer who not only wanted to hear us out, but would try to find solutions to obvious problems experienced in the antipodes.

Again, our Worldwide Service Managers' meeting had been a successful engaging of minds and many issues had been resolved. It would be a few days before I would be back in Australia, but I settled into my seat and relaxed. A major impediment in the development of the 8401 had a resolution, and now Noel's pet project could move forward.

While others in the engineering and drawing offices concentrated on getting the 8401 production ready, I knew that new owners and service manuals would be required and this was a responsibility for my division to produce. Despite thinking we could copy and paste sections from other manuals, all the effort we put into it took time. We had to edit much of the original text, rewriting some sections to keep up with the changing times

Back in the office, progress was moving well on the 8401, particularly now we had the four-pin planetary in stock and ready to go. Up until this point Detroit had no idea of the 8401's development going on in Australia.

That was until Noel, with face as dark as thunder, came into my office and quietly said, "Dearborn is now aware of the 8401 and the axle difficulties." He added, "Someone from here has given them some specific details, so I want you to find out who has leaked it and get back to me as soon as you can."

Understanding how strong he was at following company rules and that he knowingly had been operating outside his jurisdiction. As far as his contemporaries in Dearborn might be concerned, Noel was now living inside a hornet's nest.

I don't need to go into detail, but Ford has a great history of retaining records of all communications. By following the paper trail of faxes, four days later I discovered the evidence that led me to the person responsible.

A sub-contractor, he had far exceeded his authority and Noel demanded swift and appropriate action take place. An error or deliberate attempt to ingratiate himself with the engineers in Michigan had caused him to lose his job. An action of his own making, his was a role that he knew demanded extreme confidentiality.

After the unfortunate disclosure of the 8401 to Detroit's Management, I suspect that they were somewhat disturbed over the forward thinking by their Australian General Manager, Noel Howard. It didn't take long before we were made aware that there were definite rumblings arising from United States.

Noel however, was determined to proceed with the project. Taking strong moves to cement his belief to introduce this Ford tractor into the 100hp market in Australia, he rallied his allies within Ford Australia and other international operations, believing if the tractor worked in our market, it also had possibilities in other regions. I had known for a long time how strong Noel could be, and that strength became very obvious when Detroit decided to move in and deeply question the development of the 8401.

Even before the errant fax being sent to Detroit, Noel had really planned this new tractor and it had been from his memory of an earlier meeting with Company President, Lee Iacocca that gave him confidence to press on. However, it would still be important to present his idea to the CEO of Ford Australia, Brian Ingles and convince him of the profits it would bring. Knowing from experience and company folklore, that Ford will often forgive you if you stuff up, but if the same stuff up makes money, they will Lord you, Noel pressed on with his pitch and at the end of his meeting, Brian too, expressed his confidence in the plan.

By building a close association with Brian, Noel was able to enlist the local Ford Design Centre to draw up a shape for a new tractor bonnet. They designed new Ford graphics and on Noel's request designed a distinctive colour scheme to set the Aussie tractor apart from its fully imported cousins. The Australian Head Office helped by getting build costings nailed down and every step of the way Brian Ingles was kept informed of each development.

News arrived that Gordon Guthrie, a senior manager from Detroit was heading to Australia to discuss and review the 8401 project.

Even before learning Gordon was arriving, Noel had his plan ready. He knew that senior management would be looking for payback, so the moment Gordon arrived in the local office, Noel took him directly to meet Brian Ingles. Brian was supportive and after pleasantries were exchanged, he invited Gordon to view their sparkling new 8401 in the showroom of Ford's new Automotive Design Centre. After seeing the tractor in the flesh, Noel's team had arranged for Gordon to visit a local farm, where he could drive another of the prototype 8401s.

Gordon was then invited by Brian to meet with all the personnel involved in the project. With all the key people in attendance he could interrogate the design process, review the final costings and profit projections.

I was not privy to the final outcome of the meeting, however, looking at Noel's face the next day, it became patently obvious that Gordon Guthrie would be returning to Detroit full of good news. Gordon must have been persuasive as the 8401 would become the winner Noel had envisaged. However, even though Noel achieved his vision, it soon became clear that some executive members in Detroit were not at all pleased.

However, history has showed that the Ford Tractor Operations Australia delivered seven hundred and thirty-nine tractors into a falling Australian tractor market and unlike Chamberlain-John-Deere and International Harvester, we did it without the need of a bounty from government.

The design and manufacture of this unique tractor, clearly vindicated Noel's original plan.

FILLING THE ENGINE WITH OIL

I always enjoyed visiting owners, either on their farms, and at field days or at customer nights, there would always be a character who could make me laugh. On one of my first trips as a Ford Area Manager, I had to visit some dealers on a short run into rural Victoria and meeting such a character was about to occur.

This event has remained in my memory bank for good reason. It illustrated that there was always room for understanding the needs of customers who are being swept into the future without having the mental tools for them to move into a new era. Later, as service manager, it helped me focus on how our training and service manuals were written.

Arriving at the dealership and after exchanging pleasantries with one of the staff, the owner of the business asked if I would like to meet one of his customers. It sounded like a great idea. It meant I would have one on one time getting to know him, listen to his story about his community and how he fitted into the area.

We travelled for about twenty kilometres of which, the final five kilometres was slippery sand lightly covering deep corrugations. Mallee scrub tangled down each side of the track. I knew these types of roads, too fast and the suspension would dart the car everywhere, too slow and you could vibrate onto the verge and end up bogged, or crashed into a tree. However, the dealer principal knew his stuff and matched his speed to the conditions. We may have been covered in dust, but he got us there.

Driving into an open area that made up the home paddock and a smaller house yard, I could see a well-maintained lawn, a thriving fruit orchard and vegetable patch. Beyond the gardens stood an equally well maintained and very tidy shed. The Ford 5000 tractor we came to see looked in showroom condition. Parked in a bay alongside it appeared to be one of the last Sunshine Harvesters and it too had been kept in a similarly maintained order.

Getting out of the car, we were met by an elderly farmer, his handshake firm. From his lined and leather tanned face beamed a welcoming smile, I knew I liked him in that instant. For a few minutes we made small talk before chatting about his property and the new Ford 5000 that had replaced his team of horses. Inviting me to look at his prized possession, he told me that he had found some difficulties that he thought our Ford engineers needed to rectify.

As we approached the tractor he said, "Changing the engine oil is really a headache, not as bad in summer, but it does upset the wife in winter."

Frankly, I couldn't understand what he had been alluding to and wondered if his statement might be taking us into the unknown. However, reaching the 5000, I could see an open top twenty litre drum under the engine with a large amount of used oil in it. Looking up, there was another twenty-litre drum hanging by a chain from the roof, this one looked new. A small copper pipe had been soldered into the bottom of the suspended drum. That led to a tap, from there another copper line ran right to the dip stick hole.

Continuing to explain, he said, "draining the oil is fairly quick, but refilling the engine is the main problem; even in summer it takes a fair bit of time. However, it's in winter that the wife gets upset," he shrugged and looked back to the house, 'because the oil won't flow as easy when it's cold, I get it hot by leaving it on the kitchen stove for half an hour, before filling the top drum."

Trying not to erupt laughing, I told the owner that I believed I could help him. Lifting the side bonnet, I started unscrewing the wing nut on the battery tray and with a gentle tug swung the battery out of the way exposing the rocker cover. Without a word, I reached in and unscrewed the oil filler cap.

I asked him to give me a clean jam tin, which he did. Next, I scooped up some of the old oil to pour into the rocker cover.

He touched my arm, “stop! You’ll flood the cylinders,” he said.

The dealer asked him to watch as I poured the oil into the rocker cover. We all stood back, then the farmer crouched down to get a better look and within a few seconds, the oil I had poured in the top, started running out of the sump’s drain hole.

In that moment he understood, and to this day I wish I had had a camera to capture the bewildered look on the old man’s face. It was priceless, and although he had been convinced, he asked me to repeat the demonstration.

I asked him if he had the operator’s manual, he shrugged and turned to the dealer, asking him the same question. The dealer told him that they had given him one when the tractor had been delivered, describing it as having a blue and grey cover.

This time, the dealer’s statement had been met with a blank look, and the old fellow said, “after you brought the tractor out, I caught the old bull chewing something that might have been a book."

I promised the owner I would mail a new book to him as soon as I returned back to the office.

During our Friday meeting in our Broadmeadows office, I related the story to my colleagues, all of whom flatly refused to believe any part of it. I had wished for camera, which would have been the proof to the story's veracity. Today, as everyone has a camera in their mobile phone, it would be easy to provide proof.

Oh! How I wish to live again in those early days of innocent ignorance.

DEALING WITH CORPORATE BULLIES

'Old habits never change,' is a saying that had been drilled into me many years ago and is one that still holds true today.

That phrase had come to mind on many occasions, especially when arriving at the Broadmeadows tractor plant at 6.00 am. It was just like starting work on a farm. Arriving early was great, it gave me time to catch up before the office filled with many different noisy faces, along with those who came to my office bringing questions and requests.

A couple of times, I can remember arriving early and walking down the passage to my office. I would stop and look at the name on the door. Then I would ask myself, "how in the name of heaven, did you get here?"

Going in and sitting down, a stern voice inside me said, "well, you're here now, so you'd better get started."

It is quite surreal to think that a small boy from a tiny outback place like Burraboi, with basic schooling could fill the role of National Service Manager of a multinational company like Ford.

Before our annual World Service meetings, I made a habit of remembering my humble beginnings and telling myself not to take anything for granted.

To prepare, the service team and I would begin by reviewing all of the actions our department had taken over the previous year. These may have been to do with problems we encountered, or repeat warranty concerns. We would explore sources of unique failures and the frustration of a never-ending trail of paperwork. Whenever we found a way that we thought may speed things up, my report would highlight a problem and offer our solution.

These meetings in Michigan were always convivial and eagerly attended. I like to think, that through our meetings we provided manufacturing with practical solutions to some of the difficulties we faced in the field. So, every year after rehearsing my presentation, it was back to Detroit. Once there I would meet with the whole group of twenty-seven, and together with our new

Worldwide Service Manager, Bill Hepburn, we would review the past year.

Service Representatives from UK, South Africa, South America, Middle East, Australia and Detroit. Ian O'Rourke standing 3rd from right.

Prior to Bill heading the group, Larry Tarp had held the position of Worldwide Service Manager for a short term. Larry had an excellent mind, but he was tough, very tough and on many occasions I had major arguments with him. He continually accused me of having the highest base warranty cost worldwide, something for which I had no answer.

At $12.50 per hour my claim rate could not be compared with my colleagues who had been paying under $2.00 per hour for warranty in some parts of Asia.

In Australia, our standard of living was higher and this was how the basic wage had been calculated. The trickle-down effect having an impact on my hourly rate reimbursement. However, it became even more complicated where some of our dealers held an automotive dealership too.

Ford Australia's auto division was paying $16.50 per hour for customers who would bring their vehicles into the dealer's premises for service. The very same dealer's tractor servicemen would have to travel out to the owner's farm to rectify a warranty issue. At the same time, they would perform the warranty rectification without any travel time or mileage being claimable. The matter had become a bone of contention with me and I never refused to raise it as an issue.

At a later meeting in Detroit, all of the Service Managers were to be present and our discussions were slated to commence at 8.00 am. Larry would be up front.

Every so often in these meetings he would drop a sheet of paper in front of each of us, and always it would be upside down. At a specific time, we would be told to turn the sheet over. There would only be one line on the page, an impact message. Always very topical and very forceful.

Around 11.00 am, Larry asked me to stand up and after I had. He said, "Ian is the man who costs our company more money in warranty, than any other person in this room."

I can't say what boiled up in my mind at that moment, but I remember picking up one of Larry's impact message sheets that we were given earlier. Holding it aloft so everyone in the room could see, I boomed out the message in a loud and clear voice, reciting. "Never make a decision until you have all the facts." Then I sat down and felt a dreadful silence wash over the room.

The silence crushed in on our group for what seemed like minutes, but Larry never responded.

When we broke for lunch, a group of my colleagues drifted away from Larry to speak to me about my outburst. In doing that, it became apparent, that they had all agreed that he had copped what he deserved. However, it wasn't lost on me that within an organization as large as Ford, bullies would always be striving to force their way toward the top, and if they had to persecute someone to do it, they would.

After arriving home, Larry flew one of his guys out to our Campbellfield office to go through our Warranty System. He had been tasked with looking for any impropriety. However, after ten days of digging, this poor pencil pusher went home with nothing but the possibility of an adverse reaction from Larry. I still think of him sitting in his Michigan office, chewing out the accountant for not finding or fabricating anything.

However, Larry's persecution of me was the only ugly part of our Detroit meetings. These were, in my view, a roaring success. We all had a common goal to improve Ford Tractor's reputation. We swapped names and contact numbers of people who were helpful to us. Shared ideas to shortcut the company's bureaucracy and simplify communication channels. We even tabled a calendar that would enable us to keep tabs on important events. More than once, there would be occasions where our service personnel joined forces. This way we could gain much more momentum to resolve a situation, rather than coming at it through our regular or individual channels.

Before heading to the United Kingdom for our meeting, my team had reported that we had two 6610s with high oil consumption, the dealer's technicians thought the rings were of poor quality, saying they felt soft. News about them was beginning to circulate quite quickly through the area, so to put the rumour to rest, I asked that a sample of the parts be returned.

I could tell by handling them something had gone wrong with their manufacture and set about sourcing two sets of superior ring kits from

REPCO in Moorabbin, Victoria. The same day I sent them out to the dealers for fitting to each 6610.

Convinced the ring kits we held in stock were also soft, I acquired a few extra REPCO sets to 'bootleg' into our National Parts Depot Stock. To head off any repeat failures, I ordered the remaining soft ring kits to be removed from distribution.

This action had been happening while I was preparing to leave Australia to attend our Worldwide Service Meeting in the UK. By taking immediate action, our team had managed to rectify the suspect engines and put a halt to the gossip being generated by our competitors. However, I decided to table the issue for discussion and subsequent resolution.

At the meeting, I raised the fact that we had experienced an issue with two 6610 tractors with failing piston rings in Australia, informing the group that two tractors had begun using high quantities of engine oil just after delivery. I explained that our dealer technicians had reported that on removal of the piston rings they were found to be very pliable and soft. I told them that as part of my investigation, I ordered a new set of rings from our National Parts Depot and after opening the packaging, found that they were of the same questionable quality. My colleagues from New Zealand and South Africa agreed to finding similar difficulties with the same model tractors in their jurisdictions too.

John Foxwell the Senior Engineer had been listening intently and when an opportunity arose, replied,

"We had the same trouble when we were testing the prototype 6610. Believing the parts were not of suitable specification, we changed the rings to a superior version, prior to the release of the new 6610. We immediately began cancelling the original versions in our parts books and on microfiche. This means that your parts divisions should now have new piston rings in stock and they should carry a *B* suffix in their part number and definitely not an, *A*."

After explaining what I had done to manage our particular situation and that I had our Nation Parts Department remove all offending items from circulation. As a group, we all agreed that the early part is the only version available for restocking, and that nothing holding a *B* suffix showed in our records as being available.

John said, 'I'll check now, and have an answer for you after lunch."

Sure enough, after lunch he returned with another gent in tow, who John introduced as a company accountant. John explained that following the release of the new part number, his memo advising of that change is passed to our accounts for approval. Telling us that his colleague, who was standing right there, said that he wouldn't release the new parts into the system because there were still some thousands of the old rings still in stock.

It was a classic bureaucratic bungle and all because people who didn't have a mechanical understanding were promoted into positions of authority. The guy wasn't being mean, he was just ignorant of the situation and unaware of why the original rings needed scrapping.

The accountant left us after we had given him an insight into how much it would cost the company if the current stocks weren't superseded. We all felt a sense of relief when John later reported that all the old stock had been dumped on his instruction just under an hour earlier. Therefore, from that moment on, only the improved parts would be available for distribution.

This single exercise proved that the annual meeting of twenty-seven Service Managers had proved to have been a worthy idea.

Bill Hepburn also introduced a new concept to each meeting. We all had to present a twenty-minute address on some aspect of our respective areas. I felt proud to present my first subject. I commenced with the size of Australia, followed by the diversity of agriculture and the seasons we experienced across the country.

I have always been amazed at the country we live in and how Ford Australia Limited operates within this large island. For me, I have always believed it was easier for our small organization to function without World Headquarters always having their microscopic eye on us. Being so far away, our outpost had slightly less scrutiny when compared to areas like Europe or North America.

There, Head Office would always have a closer association with their local divisions. For the company's chief executives, a visit to Australia meant leaving their lofty towers in Michigan and put up with a lengthy and tiresome trip. Besides, we worked closely with the Automotive Division in Australia and they proved a valuable foil as well as being a close ally.

For anyone in Ford, no changes could be made to any product without Engineering Approval. An edict almost as old as the company itself, it had been implemented to avoid costly legal actions. Minor changes could be made by Dealers at odd times, or when the owners asked for help, but in those instances responsibility for those modifications remained with the dealer.

Sometimes and prior to an official 'fix' arriving, rules had to be broken. These occasions came whenever a new or serious failure took place in the field,

Such a case was arising with 8700/9700 models rear axle seals leaking profusely. It happened during extreme working conditions when that year's Australian cane season gained momentum. After receiving several calls from our dealers in North Queensland, I eventually obtained a set of failed seals. Our investigations proved that the originals were obviously unable to withstand the sixty-degree working conditions.

Calling around a few of my good contacts, a stronger set of seals were located and the affected dealers were verbally advised to fit them. Although confident these new parts would work, I felt relief when our dealers were soon reporting that leaking had stopped and their tractor owners were happy.

BOREHAM HOUSE

Our third Service Managers Meeting was to be a week-long event and it would be held in Detroit. The meeting would last the week, after which I would fly to the United Kingdom.

As our Detroit meeting finished on the Friday afternoon, I took advantage of my days off to fly to Orlando, Miami in order to take in a short visit to Disneyworld and NASA Space Station. It was really amazing to stand outside Runway Thirty-Nine and envisage the huge Columbia's landing there. To look at the machinery that carried out each Columbia rocket for launch was an incredible sight.

On the Sunday evening I flew to London, arriving in time for the second week's meeting. Our service group gathered at Ford's Boreham House, a place I had visited three times before, here we would attend a new tractor launch and complete a series of training sessions.

Boreham House was a stately building, three stories high and situated in expansive grounds near Chelmsford. It had been bought by Henry Ford in 1931, when after touring Great Britain, he envisaged a permanent venue for training the young farmers in the mechanical advances of agriculture. It was not an entirely philanthropic venture. Henry knew it would serve his European operations as a training and testing centre too. The main house had been designed by Scottish architect James Gibbs and Edward Shepherd the builder, commenced operations in 1728 taking five years to finish it. When I visited, the main house could accommodate up to thirty guests. The company employed enough staff to include preparation of meals and maintain the numerous outbuildings. These structures were used for housing machinery, while others were suitable for conducting training.

Besides a Farm Manager, Ford employed a permanent staff of instructors who worked there each day. The senior man was Mike Hutton, a technician who had accumulated enough experience with Ford and other products to have become a font of knowledge on mechanical farming. Mike was my go-to man on all of our company's tractor products.

Because of his experience and knowledge, Australian Tractor Operations were fortunate that on two occasions I was able to bring Mike and his expertise to Australia. A man who was ever easy to get to know, Mike rearranged his diary and agreed to assist with service training. He made an instant hit with our dealer technicians and Mike made these training sessions special for them. All accomplished with his usual aplomb.

Catering at Boreham House was under the control by a very capable lady, Mrs. Ives and to assist her, she employed a rotation of au-pair girls. In the basement, a fully stocked tavern was run by George, who had a fantastic memory for faces. It always amazed me that he could remember every visitor who had breasted his bar. As far as I was aware, this was the only place of its kind that had been provided by Ford for its junior executives and invited workers.

To reach George's domain it was necessary to negotiate a narrow descending staircase that had probably been designed for access by servants when the house was first built. Before you reached the bar you would need to negotiate a ninety-degree bend. This got you to the final entry point, a tiny doorway with an opening that even a Cornish miner would have to duck to get through. These old houses were certainly built for those who lived above stairs.

Some years into my term as Australian Service Manager, Mike and I were in the Tavern, telling yarns and lazily taking advantage of George's hospitality. During much laughter, Mike told me that he had had a batch of well-built South African service guys in for a training school on the new Fordson Dexta. On the third day of the school when they assembled in the training area, lo and behold, the Dexta was missing. Despite checking, could not be found.

When George arrived at 4.30 pm to set up the bar for the evening, standing in the middle of the basement room was an almost smiling Dexta.

Apparently, sometime during the previous night, the South African students had dismantled the whole tractor and carried it down the narrow staircase and reassembled it. How they managed to carry the major parts down the stairs and into the room is a mystery. The engine alone would weigh around 300 kilograms.

I learned later that the Canadians had found the large 'coal hold' on one side of the basement. In the early days of Boreham House, coal was used to fire the boiler and provide heating and hot water for the house. Obviously, the culprits of the missing tractor had done their homework prior to moving the heavy components into George's Tavern.

Boreham House had become renowned for practical jokes. All starting on evenings when the bar was operating. However, an ever-circulating and unproven rumour was that for the majority of the time, Mike Hutton had been the main instigator.

A rumour I remember hearing, happened one evening around ten o'çlock. An instructor arrived in the tavern dressed in tie and tails. He had come from a personnel meeting and, being suitably attired, started playing the bar's piano. A group of patrons decided that the music should be taken up to the long rectangular lake at the front of the main building. One team removed a huge door from a barn, floated it in the lake. The piano, now rescued from the bar, was deposited onto the floating door and together with the pianist, was pushed out into deeper water. It was in the middle of winter, the weather bitterly cold.

Even now, I get a happy feeling when I remember being told about the incident. In my mind, I picture a man in a dinner suit playing quite happily in the moonlight. A bystander, not wanting to miss out, strips off his clothes and starts swimming out to the floating pianist. Getting to the source if the music, he begins grabbing at the door, trying to scramble aboard. In doing this, the piano laden raft begins tilting sideways and at this point, the door begins sinking at one end. Undeterred and just as it reportedly happened on the Titanic, the musician plays on. I imagine seeing the door sinking at the swimmer's end. As it reaches forty-five degrees, in slow motion, the piano, the piano stool, the player and a want to be Bing Crosby gradually slide down into the chilly waters of Boreham Lake.

Mike had tears in his eyes as he recounted what he'd witnessed, the whole episode as he described it must have been spectacular. He went on to tell me that Boreham House's ornamental lake had more than one piano submerged amongst the European carp. However, orders coming from on high, instructed that no more pianos were to be drowned in future.

Despite these antics, the entire training team were very talented and imparted great knowledge at every school.

Mike also told me that one time he had up to twenty or more Indian servicemen attending from very remote villages across different provinces of the sub-continent. Tractors were being adopted in their regions and technicians with rudimentary skills needed training in more modern equipment.

They had come to Boreham House for training in basic hydraulic systems and they were staying there over a two-week period. As with every other training session, they were also billeted at Boreham House. However, there was a language problem and travelling with them was an interpreter who could translate their comments into perfect English.

Mike told me that he worked hard for the first three days trying to get the basics across. Then to his delight, some smiles and nodding faces began to appear. In the second week he said they were jubilant and convivial, so he knew it had come time to have some fun after they retired.

Mike said hidden in a storage cupboard on their accommodation floor was a very noisy, ancient and very unused Hoover vacuum cleaner. Tiptoeing his way to the cupboard, he took out this piece of Boreham history and plugged it into a power point, before quietly pushing into a room where three Indian Technicians were sleeping. Waiting a few minutes to ensure the men were still asleep, he switched it on.

Being old, the rusty bearings and out of balance fan had the Hoover wailing like a banshee, its noise piercing the sleeping silence. The heavy door flew open, one student in a loin cloth came bolting out and was found at the front road entrance sometime later. The second guy jumped out of the first-floor window with only a bed of roses to break his fall. Mike then heard a large bang, followed by several thumping sounds. Dust filling the room and hallway before the old Hoover screamed its last.

With a large curved sword in hand, the last technician appeared swiping at the swirling dust.

"I got him boss," he said.

He had hacked the old Hoover to pieces. The team spent quite a lot of time looking for one of the men, eventually finding him out at the main gate. First aid was then administered to the other guy by picking rose thorns out of his buttocks.

On the last Friday of any training session, all students received a certificate recognizing their efforts during their stay. It marked everyone's growth as technicians and became an important document they could use when proving their acquired knowledge.

The Indian men were filing into the waiting bus when the interpreter spoke to Mike.

"On day one you showed us a chart with oil flows going through the system. As the oil moved through the valves, it changed colour in five different places. Yesterday you loosened a pipe and oil ran out but it was the same colour, not changed. The men have asked me to ask you, why is that so?"

It was then that Mike realized he had wasted two weeks on this project. From that moment on he did his best to ensure he explained why the colours changed on oil flow charts. It was important that his students understood the legend on the chart and what each colour depicted. Colours are used to denote either changes in pressures, or direction of flow. He had also made changes to the curriculum, to make sure all students were aware of the valve function. Because of the India technicians, Mike made sure all of his instructors knew to point out, that the oil definitely does not change colour.

KEPNER TREGO TRAINING

In Australia, because of its vast size and climatic variation, we would run training schools at different locations as needed. Each year our department conducted one day Service Seminars at seven locations around Australia. These meetings informed our participants, who were mainly day to day technicians, of any important product changes and to get a feel for their issues of the day. These were no-overalls-days and our discussions focussed on covering tractor subjects. I found them to be great days and all held with a calm atmosphere, healthy lunch, plus a couple of drinks to follow.

During one of these seminars, as we moved to the question-and-answer part of the day, one of the service managers waited for me to acknowledge him. Earlier, I had noticed him rolling a pen between the tips of his fingers on each hand and at my first opportunity, I indicated for him to speak.

Pushing his chair back with his knees he stood and slowly looked at the others and then, made eye contact with me, saying,

'Why is it that our sales and parts colleagues receive rewards or gifts for doing a great job, but our poor old service guys only ever cop flak, without ever being offered so much as a slap on the back, let alone a decent reward for a job well done?

The question he'd raised had its roots in the inequality often experienced between dealer sales and service staff. It was something that had often puzzled me too, so I told the group that I would take it on board and get back to them as soon as I could work out a plan.

Ford was a great Company to work for, especially the Australian tractor team and with Noel Howard calling the shots, we knitted together. We soon became a tight squad who could cover every base. Ford Tractor Operations had always made training of its own people and its dealer staff a major priority. Noel had no time for hackers or slackers and if he found one; he would make their time at Ford Tractors very difficult.

One particular week, Noel informed me that John Henchy and I had been invited to join a group of executives for a one-week programme at Healesville in Victoria. We were to participate in a Kepner Trego course. Neither of us had heard of it at the time and brushed it off as something we would be expected to do. So, attend we would.

There were twelve participants and we all came from different sections of the company. We gathered for a friendly dinner on the Sunday evening, where we were told to be ready for a seven o'clock start. We would break at noon for an hour for lunch. The day would close at 4.00 pm, this would be time to ourselves when we could think about the sessions and freshen up. At 5.00 pm we would be expected to meet for dinner.

A far different regime to the training I experienced at Boreham House. It seemed much more regimented, because we were expected to be back for another session in the classroom from, 7.00 pm until 9 00 pm. As we were ready to leave the night's lesson, our instructors handed us a challenge, which meant up to another two hours of homework. This was becoming intense.

The Kepner Trego course was written around finding effective ways to troubleshoot any problem. I knew Noel had participated in the same thing sometime before and that if he thought the investment in our education would help us to manage our departments better, we had better do our best to take it all in. Percy Worsnop, a Ford in house training specialist, conducted the five-day show and it was brilliant.

On that Monday we were placed in three teams of four with a knotty problem to solve. Frankly, when the teams' recorded presentations were played back on the Friday morning, our final solutions were a disaster. After the morning's embarrassment, our teams were rearranged into new crews, with another monstrous problem, for which we were given only an hour to solve. The solutions found by each group was amazing. Considering our efforts on day one, it proved to me that you can't know everything. Sometimes you need to take another's opinion to get around a curly issue.

HO, HO, HO, WELCOME TO TOKYO

One Wednesday afternoon in September, about 2.45 pm, Noel walked into my office and said, “I’m talking to the States right now and they want you to be in Tokyo by Saturday. So, you’d better phone Westpac in the Glass House and organise a return flight.”

I was astounded, immediately replying, “Noel, I don’t have a Japanese visa...”

He said, “go down to the Japanese Embassy tomorrow morning and stay there till the Ambassador signs your passport.”

“Trouble is Noel,” I said, “tomorrow is Show day in Melbourne and everything is shut”

His answer was, “go down and wait there.”

So, on Thursday morning I drove down to the Japanese Embassy on St Kilda Road and walked in right on nine o’clock. I presented my passport to the lady behind the desk and asked for a visa.

She looked over her glasses and said, "Come back Monday, because as is protocol, the Ambassador will not sign a visa for anyone with less than two days' notice"

"I'll wait," I said and sat down.

She put a blank smile on her face and kept working.

Around ten o'clock the Ambassador followed by his entourage, strolled in.

A few minutes passed before the lady called me and handing over the signed visa said, "Incredible. This has never happened in all the time I have worked here."

Back in Campbellfield, as I walked into Noel's office he asked, "did you get it?"

Like the woman at the Japanese Embassy, I could not understand why the ambassador made an exception for me. Noel told me that to get the Japanese to change their protocol, he had phoned our Australian Commissioner for Trade in Canberra, telling him that I was down at the Japanese Embassy waiting for a visa. Explaining that it was important for me to enter Japan and that I needed the visa immediately. Stressing that I needed to fly to Japan the following morning. If I didn't, two hundred and fifty new tractors that were bound for Thailand and sitting on a

Japanese wharf would not be signed off. Therefore, they would not be able to sail.

The Commissioner told Noel he would fix-it and fix-it he did. No problem.

More than once during my time at Ford, the real lesson I learnt was that in business as well as life, it's not what you know, but who you know that makes a difference. Noel being able to get through to Canberra was just another example of that.

Noel explained that other than signing-off on this wharf bound consignment, the main reason for my trip to Japan would be to obtain additional information on the Ford 1700 and 1900 model tractors. These Thailand bound tractors were of a different configuration to the units we were selling in Australia. Once I had acquired the information I needed, I would then be expected to travel from Japan to Bangkok and conduct Service Training.

Shibaura Heavy Industries had been producing these small tractors for Ford under a licensing agreement.

I left Melbourne and flew to Sydney on the Friday morning, boarding a Cathay Pacific 747 aircraft. Looking at the aircraft staff chart on the cabin wall, it showed the captain's name as O'Rourke, so I gave my business card to one of the cabin crew. She soon returned, saying the captain would like to see me after take-off. Half an hour later she came back to take me up front and meet Captain O'Rourke. Within five minutes we established we had no relationship other than sharing a name, but he asked me to stay on for a chat as the 747 flew onwards. He pointed out many towns, sea names and explained much of the array of dials and displays that monitored the aircraft's functions. I was still there when a delicious meal arrived accompanied by a very large bottle of champagne. We enjoyed each other's company until the aircraft started nosing down on its approach and my new friend asked me to return to the main cabin, as he was preparing to land at Hong Kong airport.

At the airport, I was met by Peter Cook from Boreham House, he checked me into my hotel and I guess he put me to bed, as I was in less than perfect shape. It was a great couple of hours spent with my new best friend and non-relative, Captain O'Rourke.

Saturday morning, I flew off to Narita in Japan and checked into my lodgings at the Prince Taranaki Hotel. I needed all of Saturday and Sunday to recover from Captain O'Rourke's mischief the night before. It was with a pounding head I began preparing for my meeting on Monday.

I had a meeting scheduled with officials from the Shibaura Company, there we would be discussing our plans for the coming week. I was advised to take a train from Tokyo to Matsumoto City as it would be much faster than taking a plane.

A few surprises awaited me on arriving at my Matsumoto Hotel, and to add to my astonishment, I would be there for four-nights of accommodation. The hotel's front door was only a metre wide, my room on the first floor measured about three by three and a half metres. Even the bathroom was crammed, the combined shower and toilet was contained in an equally small cubicle crammed against the bed end.

From the entrance of the room to the bed head would have been little more than three metres. It was so small that my case had to be sat on the bed for me to access the clothes I needed. I then had to stand it on its end by the bed head so I could get into lay down for the night. This kind of luxury was something I had yet to acclimatize to.

Luckily, I was not a commercial traveller in that city, as I learned their lodgings were the latest thing and were more like torpedo tubes than rooms. The cylinders were about 500mm in diameter and eight of them were stacked in a group, one on top of each other. It was hard to imagine how they would be received in the

Western World as the guest required a ladder to enter them. I thought it modernity gone mad.

Dinner each night consisted of four cubes of rice about 20mm square by 120mm long and on top of each cube was the most bitter sauce I have ever tasted. The side dish was a pile of seaweed. I guess this type of Japanese cuisine is an acquired taste.

The Shibaura plant was immaculate, all production staff wore white overalls and during training white "*gruves*" had to be worn. Most Japanese have some difficulties pronouncing some English words. One of the teachers could speak English very well, so he made it easy to learn about the differences between the two versions of these small tractors. Having completed the schooling by Friday morning and after saying goodbye, I boarded the train to Tokyo for a final briefing.

Peter Cook was there when I arrived, so we decided to eat in one of the hotel's long dining rooms that were fitted with two slightly curved benches. On each side with a barbeque in the middle, seating six guests each side. As I was very hungry, and the woman taking our order had about the same knowledge of English as I did Japanese, it took some time to order. With hand gestures and broken Japanese, I explained that I wanted a large steak, some vegetables, mushrooms and definitely no seaweed. Peter ordered the same.

The chef arrived and after spending the first two minutes demonstrating how to throw two large knives up and down, he started cooking. The entire room had four barbeque areas across the room where people sat ten or twelve deep. The chef kept going from each area to the next as he finished cooking.

Peter and I felt a little out of place in shorts, and short sleeved shirts, when compared to other diners, women in intricately embroidered kimonos and Japanese executives wearing expensively tailored suits.

Two wooden mugs of beer started our repast, but were sunk on arrival of our food. Each of us had a steak 50mm round with a half a mushroom on each plate and that was it. Within a short space of time, we agreed to leave. After the shock of paying our bill at US$135.00 each, and still being hungry, we trudged down to a big yellow sign. There we paid US$13.00 each for a couple of Big Macs. It demonstrated to us, the power of American marketing at its best.

That Saturday, Shibaura's Chief Engineer had invited both of us to a company dinner. As arranged, he picked us up in his chauffeur driven car, which took us to the Shibaura Training School. This centre had been built as a huge house. It had suitable training areas, accommodation along with a beautifully appointed dining room. Peter and I were seated each end of a long table with six members from their plant staff on either side.

This time the meal and drinks were superb.

FALLING IN LOVE WITH THAILAND

The following Sunday, it was off to Bangkok, Thailand where I had arranged to catch up with George Cole. George had been Ford's Pacific Manager for tractors before moving to Thailand and joining Anglo Thai Motors as their General Manager.

He took me home to meet his wife and have dinner with them, after which I asked him to give me a full rundown regarding the setup of their Dealership. He informed me that the main business was located just near the airport and that they also operated several other branches, which were strategically scattered across the country.

He said they had arranged for me to have groups of about twenty-five staff who would come in and attend a series of three-day schools. I could expect to receive more groups of three days, for the next three weeks. I was told that they expected I would be including training that also covered new tractor and equipment sales. He also expected I could provide suitable training for their parts and service people.

This was a huge ask and something that was far more than I had been told before leaving Australia. It didn't matter, if George and Noel had arranged these conditions, I believed it had to be done.

Aware that I was in a foreign country I wanted to be sure I didn't offend anyone, so I asked George if there were any customs that I needed to observe.

"Yes," he said, "when speaking, don't shout, or raise your hands too high, never expose your palms, or point at anyone. Five o'clock is knock off time and they will all rise and leave the premises as soon as the hooter sounds. Even if you are speaking, mid-sentence, they will leave."

As I was driven home to the hotel, George's words played as if on a loop in my memory. What he said was quite different to anything I'd experienced before and I knew I would have to be very careful beginning the next morning. So, it was up bright and early for breakfast and then, it would be time for action.

My interpreter was a Thai native called Prasit, who spoke perfect English and sometimes I was taken back hearing a typical English accent coming from this warm and Oriental face. Employing two blackboards, we managed to get the training across to those attending with Prasit translating my words into Thai. Where the training required diagrams, these were drawn on the blackboards with me naming components in English and Prasit copying them across to the other board in Thai. By the end of the third session, I was beginning to wonder if Prasit could manage the training on his own.

In our first session I was feeling confident that the group were responding well and it was all going okay. I had both male and female attendees participating. Around ten o'clock I noticed two old service guys, whispering, making signs and pointing at the board. I tolerated their antics for at least ten minutes, before it was time to intervene.

Forgetting everything George had said, I spoke rather loudly, held up my hands and pointed to them. This was followed by a deadly silence and I watched dumbfounded, as talking quietly, they all moved to a rear corner of the room.

Prasit moved in close before speaking to me, "Ian, you have blown it, they are offended."

Following their pow-wow, they sat down again with a spokesman speaking to Prasit, then he began interpreting to me. "He said they understand, as you are Australian."

Trying to reclaim my position and show I understood my error, I bowed low, keeping my hands down beside my legs, when I heard clapping and voices in Thai

Prasit said, "They are very happy, now continue as they understand."

I don't remember running into any more trouble but I'm quite sure their rules were broken time and again over the next three weeks and without any more strikes. However, they did not again, act like a union.

For practical reasons there were two sheds with two groups doing work in each as I walked around. The roof line was about 1.5 metres apart. As it was monsoon season, there were downpours of rain so heavy, that the crew next door could not be seen and because of the noise all speaking ceased until the storm passed.

Late one afternoon, I was discussing different ideas to help their sales team and, in a question-and-answer forum, I was feeling great as everyone was engaged and replies were coming from most of the group. Then the damn hooter went off at 5.00 pm again. Astounded, I noticed that nobody moved and I continued talking for another two or three minutes until I said, "great day, we will continue tomorrow."

They rose out of their seats in unison and left.

George was standing at the back of the room came up and said, “congratulations, in all my years dealing with Thai staff, they do not stay once the hooter sounds. Amazing!”

Before the second week started, Peter Cook arrived to help me and assisted by taking over the practical service training sessions. A move which really lightened my load.

One Friday, Prasit said to me, “I’m taking you down to Pattaya Beach to meet the old owner of Anglo Thai Motors.”

He was a grand old English gentleman and having been born in the United Kingdom, he came out to Thailand in the 1930’s to establish the Ford tractor Dealership. His huge property had the full title to 500 metres of Thai seashore. Walking into the water I found it was quite warm and would have been great to swim in. The old man had servants looking after him and as well as the housekeeping, they would also cook wonderful meals for him too.

After lunch on a Friday my assignment was concluded and it was time to say farewell. George had planned on taking me to the airport, but I became concerned as we seemed to be going in the wrong direction. Within a few minutes we drove into a large showground area with something like a band rotunda in its centre. I was invited to join him in the covered building and I couldn't help noticing that all around the fence of the ground, were a lot of women with their children and in the crowd, I recognised many of George's staff.

George told me that the staff had asked if we could have a party to send me home. I felt so honoured to be among new friends and sitting at a table covered with exotic fruits and a never-ending supply of platters heaped with prawns and crabs. As we ate, the onlookers kept calling out my name. Lunch had almost closed when Prasit approached me carrying a model ship made completely of glass, it was beautiful.

In the centre covered section there were glass figures of the Thai monarchy. Prasit explained the figurines were models of the nine kings who had ruled Thailand and that the fifth figure was the best king ever to rule their country. He went on to explain that the Thai Motors employees were presenting the ship as a token of their appreciation. Needless to say, I was gob smacked.

I flew home that evening with my gift carefully wrapped.

One year later I was flying back to Australia from London with Thai Airlines, we were landing in Bangkok at five o'clock to change planes. After disembarking and walking into the main Reception area there must have been fifty or more of my old students and family calling out my name and chanting, “Ian has come back to see us.”

Apparently, George Cole who I had met up with at a service school at Boreham, must have phoned ahead and revealed my travel plans to them. I found it all, a very humbling experience.

TEXAS AND ITS LIQUOR LAWS

At one time around midday and following the release of a new tractor model, I was aboard another 747 and flying direct to Dallas in Texas. When I boarded the temperature was about 39 degrees Celsius. The flight was quite comfortable and after enjoying my fill of First-Class hospitality, I slept my way across the Pacific. On arrival I was met by one of the American Instructors who would drive me north to the Ford Training Farm in Paris.

When we checked into the Motel, I was asked why I was so brown. Because the outside temperature in Paris was now down to thirteen degrees and snowing heavily and I had arrived under dressed for the weather we were experiencing, I guess the receptionist was curious.

There were a few idiosyncrasies of Texan law I had to come to terms with as well. At the time, one of their more unusual laws, was that alcohol could not be purchased in a liquor store, only when eating in a restaurant. However, that may now be different in today's world.

During my time with Ford, I have found that across the whole company, there are more than a few personnel who often like a drink after work. Sometimes, and because of the liquor laws in Texas, we had to drive one of the vans about twenty-six miles into Oklahoma. There we could purchase beer.

As we crossed the border, I noticed there were old African Americans sitting on their verandas. They must have been taking advantage of the opportunity in the difference between the Texan and Oklahoma's alcohol laws. These old men would be sitting in their rocking chairs, smoking corn cob pipes and pointing to the sign saying they had cartons of beer for sale. The sign boasted, 'Beer at Special Prices'.

I wondered who was getting the benefit of the special, them, or us. It didn't matter we were after beer and after handing over the cash, we would take our cargo back to the motel. There we would borrow two rubbish bins, place our spoils inside and fill them with snow, it was a great idea and we were all ready for a bit of beer consumption later.

Our training sessions were difficult, it snowed most of the time. So, all of the work was carried out in a warm shed and after we finished, we could be found relaxing back at the motel. What could be better? We had snow chilled beer to wash the difficulties of the day away.

I was happy to learn that while I was there, Paris, Texas, was the home base for Campbell's Soups.

MEETING THE REAL PEOPLE

After spending any time away from Broadmeadows, I would arrive at the office early determined to catch up with the correspondence in my, In Tray. It always felt good to settle in and go over the action plan to monitor the progress of our team. I had made a habit of being in the office early and without any outside distractions, I could set up the day. In the still of my office, I had time to gauge our progress. In these quiet times, I sometimes caught myself thinking on a larger plane than routinely processing the immediate issues before me. I savoured these unexpected moments. In them, I had a chance to appreciate the size and complexity of this multinational monolith I worked for.

The size of Ford's internal mail service has always amazed me. For Ford's divisions in Victoria, a twice daily service delivered to all the offices. A van would leave the Glass House and travel to the design centre, Broadmeadows plant, National Parts Depot, Truck Plant, Geelong Plant, FASPAC in St Kilda Road, and the Tractor

Division at Campbellfield. At each location the driver would pick up and drop off mail from the other areas. Documents and important internal mail from each source would be delivered to each department's distribution point where incoming mail would be sorted and sent to its destination. The outgoing mail would be collected for further delivery. The mail centre became central to keeping all locations informed and saved precious time without individuals driving around doing deliveries.

In the tractor division, we regularly wrote service bulletins and my secretary, Jenny Moloney would start by typing one master copy. This was then slipped into an internal envelope for delivery to the printing department in the basement of the Glass House. This bulletin would then be printed and delivered to the mail centre for distribution. From there it would be sorted according to a distribution list before posting to relevant staff and each of our tractor dealers.

The company's printing division was also a large part of our outgoing and incoming mail section. Even letters to dealers or replies to customers were written by tractor service personnel then given to printing to produce and mail out. It was a huge and important part of the whole Ford Motor Company's operation.

On one occasion I received information that there were no operator manuals available for the last of our Fordson Major tractors, so I approached the manager of the printing department to see if he could print one hundred copies of the only manual that we had in our service office.

John Young agreed to do the job and soon after advising our Australian dealer network, I smiled after asking the parts department how many copies had been sold. My small risk had met a need and proved profitable.

Somehow, our Head Office in the USA had heard of our Australian reprint and requested a number of copies. For me, being able to help out our friends in the United States, was a great feeling.

Another part of the company that amazed me was the basement area of the Glass House. It was huge. Beside housing the printing section, it held the telephone switchboard. Here was a department so big that it processed all of the company's incoming and outgoing calls.

The mail centre handled and distributed internal and external mail along with the continual stream of bulletins and other mail outs.

A branch of the Westpac bank could also be found there. The bank had tellers who managed a retail section for company staff to use. Westpac also employed two travel personnel, these agents were charged with arranging flights and traveller's cheques for all of the company's staff needing domestic and overseas travel.

Last but not least was George. The company provided a space for George's Barber Shop. Having a gentleman's hairdresser on site meant company staff could go to George. It was a smart initiative for an image conscious company like Ford. George's service saved his clients from having to leave work for haircuts, dry cleaning and a range of small personal or toiletry items.

As usual, my morning routine would find me in the office before it began filling with people. If I had been away and knowing Noel was also an early starter, I would walk down to his office and give him a verbal report about the past week. We would discuss any events or service crises our service department had experienced in my absence, or problems I may need to manage over the coming week. These discussions would include any training I had completed, or meetings I'd attended while I was away.

One of the initiatives we had included on our action plan was to conduct Owner Evenings at different dealer towns. Not a small thing to manage, these evenings had been requested on the minutes of several previous Dealer Service Meetings. The dealers saw it as necessary to improve their sales push. Our sales office agreed, saying that it put a face to the people producing our products. While we understood the risks of inviting questions in a public forum, as a group, we saw the benefits outweighing any drawbacks.

Our basic plan was to arrange the night with the dealer principal, he would select a suitable venue and organise any catering. Appropriate products would be displayed in front of the venue. On the chosen evening invited customers were expected to arrive half an hour before an agreed start time. Our planning was designed around allowing enough time for our guests to inspect the machinery out front, before the evening commenced around eight o'clock.

After welcoming his guests and outlining the agenda, the dealer principal would introduce the Ford Tractors Australia personnel. If any other dignitaries had arrived, they too would be welcomed and receive an introduction.

With the evening underway, our roster of company personnel would deliver their presentations. The night would begin with a session by one of our sales colleagues. They would address the state of the tractor and machinery industry and the vision the company held for primary production in Australia. We would then show a film topical to the dealer's area before introducing new products. Members of my service team would then deliver a presentation with a parts and service bias.

Even though it could be seen as risky, we wanted to interact with our customers and so, a question-and-answer session had been included.

Our whole mission was to build loyalty to the dealer and our Ford products.

At the evening's conclusion a light supper would be served and our staff would mingle with the guests and engage in informal question and answer conversations. The dealer feedback from our owner nights had always been positive.

At times I was asked by members of my own staff if I had been concerned about a farmer getting up and firing off a stiff question at me. It never concerned me. I would handle it in the same way I always had. If the customer had a question of personal grief, I would advise them that as it would need special attention, I would discuss the matter with them and the dealer at the end of the meeting. Otherwise, it was important to just offer an honest answer to their question.

Another aspect of our service operation was to carry out Product Campaigns. These were issued to correct any malfunction, or defect that slipped past quality control during manufacture. These were to be completed in an efficient and timely manner. Detroit would advise each world location of the required campaign, then it was up to each Ford service area involved to include our Dealers. We would also send a letter to each owner advising them of the upgrade that would be carried out by their selling dealer and on a 'Free of Charge' basis.

Australia only issued a small number of campaigns during my time with Ford Tractors, but the largest was to fit an oil filter to the 8600/9600 Series rear transmissions. I was pleased with the initial uptake and I thought everything was going to plan. However, as we reached the closing date, about eighteen customers had failed to arrange to have the upgrade fitted with their Dealer. Ford would not have issued such an expensive campaign if their engineers didn't feel it was necessary. In this case, they had determined the upgrade was an essential move for maintaining a long and serviceable tractor life.

In order to manage these campaigns, I would write to each impacted tractor customer, sending the letters out by registered mail. I was always careful with my words, pointing out the necessity of accepting the free of charge improvement. This action had always proved positive as within a month most of the effected owners accepted the plan, arranging with their dealer to have the campaign carried out.

However, in the case of the 8600/9600 oil filter campaign, we had a reluctance by eighteen owners to comply with our request. As we were under time pressure from Head Office to complete the campaign, I sent another registered letter to these tardy owners. Thirteen responded in a positive way and agreed to get their tractors modified. That left five owners who showed no interest at all.

As a service team, we were disappointed with the lack of response from these final five owners. We had been doing whatever we could to coax these people into accepting the company's offer of an upgrade. So, acting on legal advice, I decided to send another registered letter to each owner. I began by advising them that if they failed to comply with the Company's offer within twenty-one days, there would be no future acceptance of liability by the Company.

Overall, it left me feeling disappointed as I believe only one or two of these owners respondcd.

EMPTY VESSELS ARE NOISEY

Whenever I had time available, it was always appropriate to visit Dealers and observe their operation. We could often be found discussing ways to improve the Company's support to them and their owners.

In every case involving disgruntled customers, I have always maintained that it is usually abundantly clear that whatever the cause and whoever is at fault, the failure must be fully investigated. Most of the time this could be managed with a face-to-face meeting, which included the selling, or servicing dealer. Some difficulties were of circumstances that could be solved by simple conversation, while others often involved correcting a misconception in the owner's mind. These instances were often brought about by the customer not fully reading or understanding the Operator's Manual. Misconception was a regular source of customer enquiries to our service department; many questions would be raised from day to day and most we solved easily.

One particular day, our General Manager Noel Howard came into my office and asked me if I knew the customer of a 6600 who had been complaining about its performance.

My reply was, "unfortunately, yes!" and I asked if he would like to read the file.

The manilla folder was brimming with pages of letters from him. Attached to his letters were copies of replies from ourselves along with any dealer correspondence that addressed his concerns. The gentleman had a habit of continually writing letters, all similar. A constant complaining about the dealer's service and the deficiencies of our products. Noel having met and had dealings with him some years before joining Ford, knew the client.

After skimming through the wad of paperwork, Noel passed the file back to me and said, "I think we should travel up the Hume Highway and visit him."

I agreed to arrange it and began setting up a meeting as soon as he left my office.

Unfortunately, there are times during a working career, or on personal occasions we meet people who find it quite difficult to be reasonable and open minded. I found this man to be one of those individuals. The dealer who serviced his tractor had done a remarkable job in helping him, and I too through the same dealer, had supported him on many occasions.

Noel and I arrived at the owner's large property. The vineyard staff had quite a number of tractors and equipment assembled in the main yard. All of the tractors I could see were a rainbow of different brands and of varying models. Most appeared to be several years old and looked to be much in need of service. Our quarry, the Ford 6600 was not in sight.

We knocked before entering the customer's office. After pleasantries had been exchanged, we were invited to view the 6600, in a nearby shed. The tractor was in a tidy condition but, none the less, his complaints came rolling out.

Noel looked quite surprised, while I could not believe the list of complaints that came spilling in an endless flow from the client's mouth.

A few of the things he found intolerable were regarded by us as minor, if not ridiculous issues. All were quite within the quality control acceptance from the factory. He claimed that as the slots in the heads of some bolts were not lined up, it meant our quality control must have been slack. He claimed that it was not easy to swing the battery out and that the holes in the grille were too big. With this complaint he was concerned that these holes were allowing the rubbish to get into the radiator area.

He found the seat could not be easily adjusted and that his was a problem because different staff had to move it every time another driver operated the tractor. He griped that the cabin noise was not acceptable. We explained it was within the advertised decibel range, and still, his complaints kept coming. He would just not stop.

I can't recall all of his troubles, but there was no limit to the number of unusual items of concern that kept pouring out of him. However, not once did he mention one major mechanical failing in the full fifteen minutes of his nonstop verbal outburst.

Noel and I endeavoured to calm him, but he just would not listen.

Noel must have heard enough and said, "from the information contained within our files, I believe the dealer and my Service Manager have done more than enough to satisfy your demands. And, as of now, we will not be assisting on the points you have raised, but please be assured that our warranty will continue as normal."

With that, Noel and I walked away.

As I closed my door, I clearly remember hearing disgust in Noel's words as we drove away. Looking over to me he said, "there is nothing worse than an academic with too much time on his hands."

RECOGNISING THE LITTLE GUY

As I had promised at an earlier dealer service managers meeting, I had started checking into various ways our dealer servicemen could be recognised and rewarded. These men played an important role in keeping our products going. Often only lip service was paid in recognising their abilities and by using common terminology like saying they worked at the coal face, dissolved the discussion. However, I had promised to look into it and saw it as a reasonable request.

Discussing the matter with Noel, he agreed with my basic plan of inviting dealer service staff to accept a trip to the United Kingdom for training at Boreham House. This would be a chance to attend a formal dinner and between the prescribed activities, they would have an opportunity to spend some time for rest or sightseeing while in England.

A letter outlining my basic plan and inviting expressions of interest was mailed to all of our dealers. A guideline cost of A$1750.00 had been set and we believed that from this, we would be able to ascertain who would be interested. After the mailing went out, I began obtaining final quotes from various travel agencies, asking for an itinerary of eighteen days with a sliding scale of costs, depending on the final number of participants.

As the interest in the trip grew, my service team began phoning their dealers to finalise numbers. By the close off date, we had thirty-two firm replies and all agreeing to an inclusive cost of $1,750 per person. Buoyed by this outcome, it was now up to us to negotiate with our travel partners to ensure we had raised enough funding to cover the bulk of the trip

British Airways won the contract. Our party was to be broken into two groups, with one leaving from Tullamarine and the second group would leave from Sydney. The flights would leave both terminals around the same time of day at the end of January 1980. Our carrier proved exceptional hosts. They invited all of our passengers, along with their farewell parties to meet at each respective airport, prior to departure to enjoy a morning tea in their lounge.

A week before departure Noel, Keith Pincott, and I went over the final costings and itinerary, making sure that all those involved would be informed of our plans. Keith was one of our Zone Managers and he was coming along to help me chaperone this wild band of wide-eyed Aussie mechanics around Great Britain.

As Keith and I went to leave Noel's office, he said, "Ian, when you get your tickets, you'd better draw another five thousand pounds."

Before I could say anything, he must have seen the quizzical look on my face and answered the question going through my mind.

"You might need it to bail a couple or even more of them out of gaol."

He regaled us with a story about a time when he took a group of South African dealers to the UK. He said that after a few of them had gotten themselves into a spot of bother with the police, it cost more than a few pounds to fix.

"I know most of them and don't believe anyone will cause a problem," I replied.

However, on his insistence we drew the extra cash. Thankfully my prediction proved correct and the trip went off smoothly.

Ian Blow, representing Ford Tractor Operations Australia attended the function in Melbourne and gave a rousing speech about the importance of a tractor's dependability. Saying that when it came to gaining repeat sales, Ford's reputation for reliability was all due to the constant efforts made by our service technicians.

He finished by thanking them for their loyalty and wished everyone bon voyage.

Noel had flown to Sydney to emphasise how important our service reputation was to the company and wish our northern dealer technicians a great trip. This trip was a first on many fronts and I found myself brimming with pride when Noel told me he wanted to address the group in Sydney. I imagined it was an affirmation of my standing within the dealer network and also within the Ford Motor Company too.

When our group from Tullamarine joined with those who had gathered in Sydney, I found myself among a very happy gathering of Ford tractor technicians. The happiness quotient kicked up a notch when the Flight Service Director advised us, that they had chosen to place us all on our own in the rear end of the aircraft. Other travellers stared at us as this news brought a wild cheer of hoots and whistles rising over the hum of conversation at our end of the lounge. The group appeared quiet after take-off and took advantage of the free drinks being offered. Unfortunately, that element of British Airway's hospitality ran out before we passed over Alice Springs. Knowing that our flight would be stopping off at Singapore our Ford tractor boys hatched a plan for spending their time and money in the airport bar.

Suitably lubricated, and with everyone armed with an appropriate allocation of *travellers*, our team boarded. At the last minute, everyone watched the flight attendants go through the safety routines before we settled in for the long flight to the UK.

We landed at Heathrow on Friday, and as scheduled, a coach was waiting. Its driver was a chap called Ken and he was accompanied by a lady called Cherry. Cherry would act as our Guide whenever it was required.

It was early morning and we were being driven to our hotel in the heart of London. I decided to address the troops, saying that after checking in, it would be a wise idea for everyone to have a rest until four o'clock. Suggesting we should meet in a large lounge and run through the whole eighteen-day programme. We waited in the lobby, chatting to different people and getting better acquainted while everyone checked in.

Contrary to my usual practice and to keep costs down, for this trip, Keith and I were sharing a room. After taking a shower to freshen up, Keith suggested that we should go for a quick walk and explore our surroundings. Being the middle of a grey London winter, it was a little on the brisk side for window shopping. Even so, off we strolled. We spent around an hour of gazing into shops and going into Harrods. It soon became apparent, that much coin was going to be needed if we were to purchase tokens of affection from Harrods to take home.

Deliberately we had planned a route that had been intended to avoid passing by our hotel. Having stayed there before, I knew the layout of the area, so there was little chance of us getting lost.

Keith said he was feeling peckish and I told him about a nearby pub that served a real British sandwich. It was only about four hundred metres from our hotel, so we made haste toward it. We were wanting to enjoy a little time away from our charges, but as we entered a rear parlour, we found ourselves in the company of twenty-four rowdy Aussie tractor men. Our companions, not taking notice of my suggestion to rest, were now in an English pub enjoying an ale or two, cheering and singing up a storm. Happy, but in a very poor state.

On Saturday after a hearty breakfast, we all piled into Ken's coach, Cherry explaining the history and importance of the sights we were passing, while we toured around London. Occasionally stopping, he would allow us to alight and move around to get a better view of various places of special significance.

Unbeknown to our party, Noel had arranged for three senior Ford Managers to attend a dinner in our honour at the Hotel at six o'clock. So, before departing the coach for the last time that day, I advised the team that the company was shouting us all a night to remember. I implored everyone to stay sober, be sure they dressed for dinner and asked them to make their way to the dining room in plenty of time.

On arrival, Keith and I introduced our guests, extending our hospitality, some beverages were partaken. On the bar I placed five hundred pounds, alas that diminished inside thirty minutes, so back onto the bar went another two hundred, which also disappeared in short time. It was now time to eat and we were summoned to be seated.

At the end of dinner, our guests from Ford Tractor Operations spoke on various topics and told our touring party about a few different agendas the Company had for introducing new products. They also told us about planned improvements they had coming on existing models.

Sunday arrived and again the coach with Ken at the helm and Cherry as our tour guide, they took us to other different sites of interest. We would not be returning to the hotel, so before we could move off people milled around making sure their bag was on board. Everything stowed away, we found our seats and listened as Cherry took to the microphone.

Our sojourn would end with an hour's trip to Boreham House where we would spend the week of training. I found the hardest part of wrangling these tourists on our trip, was keeping track of Pinky Williams from Leeton. Each time the coach stopped for the troops to take photos; Pinky never returned back to the bus on time. I could see it was troubling Ken and to overcome the problem, I put a "Minder" on Pinky. Problem solved.

The last coach stop for the day was at our accommodation, a hotel close to Boreham House where we would have breakfast and dinner. Lunch would be provided at Boreham House by Mrs. Ives and her staff.

At the end of the week our group had enjoyed a solid week of expert training and were waxing lyrical about the benefits of the whole programme, suggesting it should be bi-annual, if not an annual exercise. To say they were enthused by the training side of the trip would be understating their enjoyment. It felt good to see something that had been suggested in a service meeting had become a reality and I was proud that the event my team and I had planned had been received so well. With service men like these on our side, I knew Ford Tractors Australia had a solid future.

Friday night marked a good end to the week. We were all up bright and early on Saturday morning so Ken and his trusty coach could tour us down toward the south coast of England. Our favourite tour guide Cherry continued running a commentary as we rolled along. It was a really stormy day and at Brighton Beach. Wind pushed against us as we braved the weather to get closer to the water. Waves were crashing and white foam was building over the black roundish pebbles that were a Pommy version of a beach. As Aussies, we found it hard to get our heads around a beach without rolling surf and an expanse of white sand.

Next day we moved onto visit a museum featuring classic cars. Anything mechanical will keep a tractor man occupied for hours, but this place was a sensory overload and it took some time and coercion from Cherry to get everyone back on the coach

Monday and while Ken drove us north, Cherry announced that we were going to view "Bill's Place." She went on to explain this was the house that had been home to William Shakespeare. During the trip Ken stopped at different places where Cherry described their significance before allowing us to get off and take photos. I stayed in my seat, imagining that a year ago, very few of these men would never have thought they would be here and seeing any of these places of scenic or historical importance.

On Tuesday, we started early again as Ken was driving us to Stonehenge. This very important site and its history was being detailed by Cherry. She explained that the current thinking of academics trying to understand the reasons behind Neolithic man's need to erect these unusual and massive rocks.

Wednesday arrived with a journey back to the Hotel near Boreham, where Keith and I told our travellers that they had two days to do as they wished to, but insisting they be back by Saturday and ready for a long flight home to Australia. And, I made a point of saying, we would be leaving with, or without, Pinky Williams.

The touring party SSP Tour 1980:

UNDERSTANDING THE PROBLEM

Our office often received field complaints, either through our dealers, or by letter from an owner and a number of them stick in my mind. Noel came to me on a Friday, telling me that he had just spoken to a good dealer who had a problem. This chap was one of our top dealers who always sold a high quantity of our Ford tractor products. The problem perplexing him was that he had an owner with a 7710 tractor who had been reporting that the machine had faulty, bordering on dangerous, brakes. Even after the dealer's servicemen had replaced the components twice, the operator was still reporting that he could not get it to stop.

I told Noel that I found it hard to believe, as across all of our models with oil enclosed disc brakes, there had until now been no reported difficulties. Checking our computer records, sales of brake parts were very low. Considering the number of units working in the field without brake concerns, I phoned the dealer and asked where the tractor was.

"It's in a thousand bits in the workshop," he said.

"Put it all back together," I said, "and I will be on your door step just after eight o'clock on Monday."

As planned, I arrived at the dealership early on the following Monday. After making myself known to the servicemen, I mounted the tractor and drove it down the dirt road running behind the premises. With as much pressure as I could exert, I stood on the brake pedal bringing the 7710 to a sharp stop. Turning round, driving back, standing on the brake pedal again provided the same result. To me there was nothing wrong, this tractor was dragging the smoking rear tyres to a dramatic but very controlled and quick stop.

My next move was to ask to see the unit in operation. Watching it working in the field was the only way I could see us moving forward and finding a resolution to the owner's problem. I asked the dealer where the owner lived, only to learn his farm was about five kilometre's out of town. With the new parts fitted, this tractor's brakes were now in the same condition as when it left the factory. Therefore, our next option was to send the tractor back to the farmer and watch it in operation. I left the dealer to make the calls. He came back a few minutes later, saying he had arranged for us to meet the owner at two o'clock.

I was introduced to the farmer and looking about, I saw a bubbling crowd of thirty or more milling around. Among them was the client's solicitor, complete with legal pad, and poised pen at the ready. Among the group was Ian Rice, the manager of their local Cane Growers Co-operative.

I asked the owner what he had been using the tractor for, when he experienced trouble with the brakes. He pointed to a row of ten fully laden mobile cane bins, to which he had attached the tractor. I couldn't believe what I was looking at.

With all of us bearing witness, the owner had his little 7710, and weighing in at around six tonnes, hitched up to ten cane bins weighing 2.5 tonnes each. He was intending to haul this train of wagons five kilometres into the mill. The load totalled twenty-five tonnes and was highly illegal. The State, had with safety in mind, legislated that the mass being towed, must not exceed the mass of the prime mover.

I took the owner to the back of the tractor and in a voice loud enough for all to hear said, "if you hit a car, or school bus travelling on your road into town and through the crash, end up killing someone? It's you who will be liable."

He spun around and looking back and forth from me to his solicitor, shrugged. With a blank look on his face, he fixed on the solicitor again and turned his palms up in a questioning gesture.

Acknowledging his client's plea, the solicitor said, "he has a strong point."

Aside from some uneasy shuffling of boots dragging uninteresting circles in the dirt, not one of the onlookers made a sound.

I then told those involved that they would receive a letter from me in a few days. I would be advising them of the unlawful actions I had almost witnessed, and to help them understand their obligations, I would enclose a copy of the relevant laws covering braking systems.

The case solved, I was pleased to report my findings and subsequent action to our General Manager on my return to the office.

This situation clearly demonstrated how constant pressure on a dealer or their staff, can cause a poor analysis when trying to resolve an owner's complaint. In such situations, bad news always travels fast, but good news usually gets minimum attention. In this case, news like this could have travelled like a bushfire with headlines in rural papers declaring that, "Ford has major brake problems with their tractors." They would have been based on a lie, but devastating nonetheless and tractor sales would have been adversely affected.

Our investigation and final result averted such a situation and it proved to me once again, that a technician needs to cut through the emotion if they are to find the real source of any recurring problem.

Looking into the whole history of the brake problems with this particular tractor, I found that the dealer had replaced all the major brake parts TWICE and had made no improvement in stopping the tractor. Surely that should have rung alarm bells, but because of the owner's perceived standing in his community and the pressure on the dealer principal to solve the issue, his technicians had overlooked the way the machine was being operated.

This was not the first, nor was it to be the last time that I had encountered a situation where there had been insufficient analysis of the customer's complaint. In quite a few cases, there was confusion in the reports being submitted with up to three variations of the owner's troubles. Every time it could be sheeted back to a poor understanding, or a lack of attention to detail by the personnel involved.

To counter this and to use in future service training, I wrote a thirty second story, then made a tape recording of it that we could use to test our dealer staff's ability for careful listening.

The recording sounded like this:

> *"At Heathrow Airport, in the UK, pilot Bill Smith and first officer Jack Thomas boarded their Qantas 747, preparing to fly two-hundred and eighty-nine passengers back to Australia. The cabin crew consisted of ten flight attendants and Brian Brown, the flight service director.*
>
> *As the 747 approached Athens, one of the passengers became quite sick, so attendant Lucy Cole attended to the distressed passenger, while the 747 continued on its merry way to Sydney without further troubles."*

Over the next series of Service Meetings, I addressed each group, speaking about the importance of listening carefully. I wanted them to understand how to ask the right questions and then how to rate the importance of any answers given, during a product problem discussion. To me this was vital for getting to the root of their customer's problem. I wanted them to understand how observing any change to their client's body language, would give clues to any fabrication or embellishment they might be making to their claims.

After finishing my introduction, I usually asked for three volunteers, keeping one of the volunteers beside me and ushering the two others outside.

I would recount the basic plan for everyone to listen carefully at all times and switched on the recorded message. When it finished, I told the man beside me to take the second serviceman out of earshot of the third, repeat what he had heard and return to his seat. Then the second man would be asked to recall his version of the story to the third volunteer before coming back in.

Again, I would go through my listen carefully spiel, and then ask the second man to repeat what he had heard. This process was repeated again with the third volunteer, again asking him to repeat what he had just heard from volunteer two.

The last story repeated was always nothing short of incredible, on most occasions, it was more than inaccurate and a total mess, proving that it is difficult to rely on memory or a third party's hearsay.

By adapting the old childhood game of Chinese Whispers, this was my way of making sure the truth of a problem came from the source and, our accuracy for describing a problem began improving.

OIL DON'T DRIP FROM THE BREATHER

On another occasion a dealer phoned me with a major situation with two brothers who had purchased two 7710 tractors and were complaining of oil dripping from the breather pipe attached to the rocker cover. The dealer was pleading with me to visit and sort it out because, apart from being good customers, they were very upset and demanding he replace their tractors with new ones.

Arriving at the dealership at nine o'clock, I found myself walking into a building that appeared to be deserted.

Asking where everyone was, the receptionist said, "John is in the training room with the owners."

Opening the training room door, I could hear a murmur coming from thirty or more men who were sitting in chairs facing the dealer. He was sitting behind a desk, holding the fort with three of his staff. Opening the door, the noise died down.

I asked John, "who are all these people and, where are the owners?"

John pointed to two men in the front row as the owners and explained that the rest were their relatives and friends.

At that stage, I held the door open and said to the owners, "would you gentlemen please stay here, but I ask that your relatives and mates go home. Only then will I discuss your issues."

Knowing that it is an impossibility to try and resolve any trouble when the room is mixed with emotion and different agendas, particularly among numerous onlookers.

I retreated to the lunch room, made a coffee and waited until the crowd of relatives had left the building.

Returning to the training room, I asked one of the brothers to tell me his story and inquired where the tractors were.

I was led into the workshop where the units were waiting. One of which was attached to a dynamometer. Starting the tractor and placing a load on it through the dyno, a small drip of oil emitted from the breather pipe every four to five minutes.

I thanked them for bringing this concern to our notice, asked them to go home and suggested they come back after lunch tomorrow.

The dealer's Service Manager seemed bemused and asked what I was going to do.?

I told him to go to the local steel shop and purchase a length of tubing fifty percent bigger in diameter to the current breather pipe. I asked him to duplicate the current pipe's shape, paint it blue and to replace the original breather with it.

Arriving back in the workshop next morning, I found a new pipe was fitted to each of the offending tractors. The one we had seen the day before was still fitted to the dynamometer. We started the engine and brought it to operating temperature before going to full power, we ran it this way for almost ten minutes without any oil dripping from the breather. We hung an empty can from it, so that it was hanging under the vent. The tractor kept running and we still saw no oil dripping into the can.

After lunch both of our owners arrived to watch our trial take place on each of their tractors.

Finally, with the owners, the dealer principal and a gaggle of service technicians watching, we ran another demonstration. Again, no oil was seen dripping from the breather of either tractor. Convinced with the explanation I had given them;

the owners were all laughing and shaking hands around our own group of smiling dealer staff.

The cause of the negligible oil loss could be put down to a design feature. Because the rocker cover is on top of the engine, there is always a quantity of oil in a suspended state and this is necessary to lubricate the valve train. While the engine operates, an amount of air pressure is built up within the engine's internal atmosphere. To manage pressure differentials within the different areas of the engine's housings, a series of baffles trap this suspended oil, returning it to the crankcase sump. This allows the separated air to exit over the baffles and into the breather pipe. As the breather exits the top and runs down the side of the engine, air escaping sometimes tends to condense in the tube, dragging a minute amount of oil particles out with it. Therefore, by doubling the breather pipe diameter we slowed down the speed of the air escaping the crankcase and there was no visible oil loss.

On another field visit to a dealer, again arriving at 9.00am, I was chatting to him about warranty and other issues when he said he had an owner coming in at 10.00am to speak to me about his TW10. I could see the tractor in the workshop and nodding toward it, asked what is wrong. Suggesting it was easier to show me than explain it, we walked over to it. The seat and transmission cover had been removed exposing the differential gears that were now a light colour of blue.

When the owner arrived, he forcefully explained that in his view the company should replace the tractor or at the very least fit a new transmission.

I asked him what had happened to cause the obvious overheating.

He explained that he had two properties about five kilometres apart. As he was working on the second farm when he noticed the hour meter had reached 1200 hours and he remembered that the manual stated the oil must be drained. So, he drove the tractor over beside the creek, then removed the transmission drain plug. When all the oil had drained out, he got out a clean bucket and filled it with clean water from the creek, pouring it into the transmission via the fill hole until it was filled, then drove home, slowly.

I asked if he had heard any noises on the way home and said he did hear some as he was nearing home. I asked him did he read the Operator Manual, to which he said not lately. The Dealer had one, so on opening it at the maintenance section to see the words "drain and refill"

But he said he believed that the clean water should have been okay until he reached home.

I didn't receive a response when I asked him why he didn't drive home as the short distance would have only added a few minutes onto the 1,200 hours as such a minute increase in time would not have caused him or the tractor any trouble.

I pointed out that water is a wonderful product that can transfer heat and does not normally compress, but stressed that it certainly does not lubricate. I asked him to look at the gears, suggesting they are usually a colour like stainless steel, but not blue.

He did not sound convinced, so I suggested that I would take a gear with me for our metallurgists to test, but unconvinced, he rejected my offer saying that they would just report what I'd want them to say.

Becoming frustrated with his capacity to understand, I explained that those guys have a certificate on their wall, advising their qualifications, which is their method of providing a true answer for an engineering failure. After much more discussion he and the dealer agreed to repair the transmission at a reasonable cost.

Sometimes it is difficult to understand how such strange circumstances arise, possibly it comes from hope or a cry of help to pry them from an unfortunate misconception.

DETERMINING LEGAL RESPONSIBILITY

Only two cases of legal action were ever launched against the company during my tenure. The first was quite tragic in one sense, however, it proves the absolute need to operate machinery with safety in mind at all times.

Two farmer's living side by side; where Farmer One had a Ford 4000 with a manual transmission and Farmer Two had a Ford 4000 with a Select-o-Speed, (shift on the go) transmission. Farmer One was about to bale some hay so his neighbour suggested that he should borrow the tractor with the S-o-S transmission that would handle the job faster, as gears could be changed according to the windrow size. According to the information given to me at a later time, the farmer using his neighbour's tractor only received basic instruction on using the S-o-S transmission.

We were advised that the farmer commenced using his neighbour's tractor when a blockage occurred in the pick-up of the baler. The driver alighted from the tractor, leaving it still running at 1,2000 RPM, the PTO engaged and the baler operating.

As the driver began to dislodge the blockage, the tractor jumped into reverse, rolling across him causing serious injuries.

As it is legal custom, a passage of time is required to establish the state of the victim's health or injuries.

Some five or six years had passed prior to the case coming into our hands. Ford has Product Liability Insurance to cover this type of action, and this involved me meeting with our insurer's representative and analyse our position and structure a defence to the plaintiff's claim of poor design.

Our first action was to view the tractor. This was located in a National Park that we had to visit. On arrival we noted that the battery was missing,

along with the Select-o-Speed gear lever assembly and cable.

We had been advised that the tractor had changed owners three times since the accident and was now in a very dilapidated state. The Select-o-Speed gear control was in the hands of an engineer engaged in the case by the tractor driver.

Heading to Brisbane, we needed to view the suspect gear control. The plaintiff's engineer allowed us to take photos, but not to take the parts away. The cable, tubing, cam and lever were extremely rusty, making it almost impossible to operate.

It was obvious that on the day of the accident the control on the slotted cam had been had been moved by the driver to an incorrect position whereby the roller was on top of the two slots between reverse and 1st gear. Vibration of the running machine caused the tractor to jump into reverse gear.

Further investigation revealed the tractor had not been serviced for some years.

We spent many hours gathering information to a point where I was advised by the legal team that we had a strong case to present to the court.

Twelve months prior to the court hearing the case, we received advice that the owner had passed away. However, the plaintiff's team were determined that case would continue.

On the morning of the case, the Insurance Company agreed to settle out of court for an undisclosed figure.

In hindsight, I still feel very sorry for the tragic accident and sorrow for the operator's family, but the tragedy also serves as a grim reminder, that when operating machinery safety is a must at all times. And, every Ford Tractor Operating Manual has explicit instructions for operating the product in a safe manner.

ANOTHER LEGAL CASE

The second legal case involved a driver working for a large Metropolitan Council who sued the Council and Ford for injuries sustained when alighting from the cabin fitted to a Ford Industrial Tractor.

In our defence, I pointed out that the cabin was not of Ford manufacture and had been a unit purchased and fitted by the Dealer. Also, one handle was missing from the cabin that would normally assist a driver to exit the cabin.

Surprisingly, their Lawyer suggested that the step on the Ford was of poor design. I produced photos of similar competitive units whose steps were of a lesser area for boots to stand on than the Ford unit

I was asked how do you expect a driver to step out of a cabin, to which I demonstrated walking backwards, holding onto the remaining single cabin handle. The lawyer claimed that in a test in the council yard, all drivers stepped down in a forward motion. My comment in such a move a driver cannot extend his arms backwards at that angle as the driver cannot hold onto the two cab handles it is not possible or safe.

After some time, Ford was cleared of any responsibility, and Workers Insurance, together with the selling dealer had to pay some compensation.

I learned some weeks later that the same complainant had worked for another council before and had won a case for injury stepping down from a tractor cabin in an almost identical situation as described earlier.

The most complicated case that I was involved in, was quite bizarre. A customer had purchased a Ford TW20 tractor only a few days after release, but following delivery, made strong complaints to the Dealer regarding numerous faults with it. Irritated, the owner contacted Noel by fax insisting that he wanted his tractor replaced with another TW20.

Noel was quite perplexed when he called me in to his office to discuss the matter. As there was nothing I could do from Melbourne, Noel and I agreed I would need to meet the man to investigate and hopefully lay his demands to rest.

When I contacted the Dealer, he insisted the tractor was in perfect shape. He had sent out two servicemen out to check over all the owner's concerns. After they had inspected the tractor on the owner's farm, they bought back a sheet noting everything that had been checked, but the owner had refused to sign the document.

The fax machine in our office continued to print out messages from the owner and after a couple of weeks, the fax machines in Detroit also started printing out reams of paper with a barrage of complaints from the owner.

Eventually, a plan was drawn up outside normal Company Policy to provide a new tractor, as there did not appear to be a suitable alternative.

A new tractor was dispatched to the dealer who was to carry out apre delivery service and then it was to be taken to the owner's property. The new tractor was parked beside the original unit in readiness for a local company to swap the tyres between the two tractors the following day, as we had agreed. Both units were parked about five-hundred metres from the owner's house.

Next morning as agreed, the tyre fitters arrived at 8.00am ready to change the tyres and were more than surprised to discover both tractors were missing. Detectives were called. However, while there were tyre marks and evidence of a large truck having been backed up to the site, the owner's statement was either misleading, or of little use. He claimed he had decided to go to bed early and had neither seen nor heard anything. To me and my colleagues the whole event seemed fishy, how can two tractors be moved without anyone seeing or hearing anything?

Independent of the police investigation, Ford decided to engage in a private investigation firm to gather as much evidence and testimony in an effort to help the Queensland Police to try and solve the obvious theft. Further complicating the situation, the customer's original tractor was under an Hire Purchase Agreement, so that company also became involved.

Despite all the efforts that were made and fifteen months passing by, finding the stolen goods proved fruitless.

Eventually the police contacted our office with news that the tractors had been found and I went to inspect the two machines. Apparently, two bushwalkers had been strolling through a thick rainforest on private property before coming across an expanse of dark green shade-cloth style mesh draped over two large shapes. Some effort had gone into hiding these machines as the covering had been anchored down by numerous steel pegs.

Curiosity must have driven the bush walkers to begin lifting up the mesh, to get a better view of what it was hiding. It must have excited them to find that there were two blue tractors crouching below it. Luckily, they were aware that the machines may have been stolen and notified the police.

Receiving information our missing tractors had been found, I made my way to Queensland to inspect them. Both machines were now in a shocking state of decay. The humidity had taken its toll, the seats in both cabins had become soggy and the vinyl covers had completely perished. All of the tyres had become flat and the perish lines ensured they were beyond any serviceable use. Touching the rusted steel radiator mesh, it fell to the ground in a thousand pieces. All of the electrical wiring was totally cracked and the covering gone revealing corroded copper wire throughout the many looms. Frankly, both tractors were nothing more than scrap material.

All the investigations had yielded nothing.

The ever-complaining owner of the original tractor also had an earth moving business and unfortunately had become the victim of a bad workplace accident only two months before the missing tractors' discovery. A tree had fallen on his 'dozer, which caused him to die from his injuries.

We learned through the investigation, that it was alleged that the property where the tractors were found had been owned by a close associate of the complainant. An event that stays within my memory, even after dissecting all of the information provided to Ford and considering the final outcome, none of us could not come up with a tangible reason as to why anyone would plan the whole saga.

A LITTLE BIT OF SUGAR ALWAYS HELPS

On another occasion, while walking down the passage at Broadmeadows office I bumped into Ian Blow, who asked me to join him in his office. As I sat down, Noel followed me in and joined us. Ian explained he was quite concerned over a lack of sales of the Ford Wheel loaders in Western Australia. From day one the loaders had been selling well in other markets, with no major service concerns.

Ian told us he had been speaking with the WA Industrial Manager who had told him, that due to the increasing number of service problems, they couldn't sell any more loaders. Noel and I were shocked, we were not seeing an undue number of complaints coming in from the other sales regions, so it was agreed that I should visit WA and endeavour to sort out the situation.

I flew into Perth on a Sunday and as arranged had dinner with John Henchy, who was at that time, our Ford Tractors Manager for Western Australia. John told me he knew nothing about the lack of sales of the loaders.

On Monday morning John and I met with our Industrial Dealer, his sales staff, and their service manager. During the meeting it became apparent there was some confusion about the situation, so I suggested that we arrange to meet a few owners in the Perth region, where it seemed to me, that this was where these rumours were coming from.

At the first property we met an owner who was a happy guy. We asked him how he liked his A66. He had nothing but praise for his machine.

I said to him, “you have a sand bucket fitted, and are using it in your quarry,” and because it didn’t make sense to me, I asked, “why?”

He explained that the multipurpose bucket had met his early needs. He pointed to some welding with his foot and went on to tell us, that when it came to quarry work, he had strengthened the original bucket by welding in some bracing straps. I could see he was proud of his ingenuity, because his remedy had fixed his problem.

I asked him how much it cost.

He said, “probably a hundred bucks or so.”

I thought about what he told me.

“To help you out,” I said, “the next time you need parts, oil, or filters for your loader, make sure you go to the dealer to get them. As a matter of goodwill, I will arrange for a $250 credit to be paid into your account." I shook his hand and winked, "it's not much, just a little bit of sugar to help cover a few of your purchases."

Our next call revealed another customer with a similar story to the first one. He was quite happy to chat about his machine, except for engine oil leaving a stain running down the side of the block whenever he had been removing the filter. This was an easy fix, as on his model there was no guard fitted around the lower section of the filter base.

I told him that when he was next in the dealer's parts department, that they would have a guard to fit below the filter for him. This modification shifted the flow of filter oil away from the engine and could easily be caught in a bucket during the change. I also promised that there would be a pack of filters and a twenty-litre drum of engine oil waiting for him as a gesture of good will.

We visited two more owners and resolved their concerns, which in my opinion were minor. I left some instructions with the dealer to supply these clients with a similar parts and filter pack free of charge to these customers too.

I felt our trip had been successful and our presence warmly received. Our reception got even better after inviting all of the people we had visited to join us at the hotel after work. There yarns were swapped and laughter shared as we enjoyed our share of, *alcoholic vitamins*.

Returning to our Campbellfield office on Friday, I presented my findings to Ian Blow and Noel. I outlined what I thought the cause of the main problem affecting our sales of wheel loaders was.

As I took them through each customer visit, Ian's secretary brought in the morning's mail. In it were included two new orders for wheel loaders from Western Australia. For me it proved that a little public relations combined with good customer service equalled more sales. These were happy days.

TOURIST OR TRAVEL GUIDE

Not long after arriving home from our tour to the United Kingdom, the tractor office began receiving calls and letters pressing an argument that there be a next trip, and for it to be to the United States of America. Speaking to Noel he supported the proposal, so I found time to sit down and put together a draft for a trip lasting about thirty days. This time we would open it up to include our dealers, their staff and their wives. Building on the success of the first trip, I proposed that we also invite our owners and their wives.

It was important to devise a plan that would include a mixture of visits to various Ford tractor dealerships, along with several different types of farming operations and outings to cities which would cater for wives travelling with us.

To justify some cost as being for training, the first draft itinerary looked something like this:

- Fly from Australia to Hawaii where we would stop over to rest and recuperate from the flight for two days.

- Next stop, San Francisco; three days visiting dealers and farms. If the ladies on the trip wanted, we would offer shopping and sightseeing tours.
- On to Seattle where we would tour the Boeing aircraft facility.
- Detroit for visits to the Ford Museum, Greenfield Village. Meet Ford executives for dinner before visiting different Ford Assembly Plants. Two days, would be put aside to visit the Renaissance Centre built by Ford and give our guests a little free time to explore on their own.
- Travel to Orlando for three days where we would spend our time visiting Disneyworld and exploring the NASA Space Facility.
- Then it would be onto a Ford Training farm for two days for those who wished. To cater for the ladies in our group, I had asked the travel agent to send me details of tours that they thought we might find interesting.
- Onto Las Vegas for another two days of exploring casinos and a trip down the Grand Canyon.
- Finally, we would fly back to Hawaii for our last two days and then suitably relaxed, we would head home to Australia.

The final draft prepared. I circulated it to our dealer network and rejoiced when I found out we had seventy initial applicants. However, joy turned to disappointment when this trip was put on hold. At first, I didn't understand Noel's reluctance and it only became clear when we were made aware of big moves happening in Detroit that were to decide the Company's future.

I remain convinced that it was Ford's purchase of New Holland that put a stop to my plan to lead a dealer and owner visit to the USA.

International travel and the planning that went into those meetings, may have formed a large part of my role as National Service Manager. However, these trips also meant I had some free time. Free time that offered me an opportunity to explore foreign lands, which otherwise would have been unavailable to me.

One free weekend in Detroit, I found time to visit some of the city's attractions familiar to visiting tourists. My first stop was the Henry Ford Museum and walking through the vast display I couldn't help comparing the space to being similar to that of about five aircraft hangars. The display was so vast and as the brochure stated, its mission to preserve America's heritage abounded. The variety and scope of the display was beyond incredible.

While I expected to see plenty of Ford products being displayed, it was the household items like radios and televisions from every period that garnered my interest. Works of art would identify a period and I would be taken back to my boyhood, remembering these things being advertised in Dad's old magazines. Differing emotions would bubble inside me as I inspected a farm truck, or be blown away by a steam-train. Often these were things I had only seen in cowboy movies when we went to a picture night in Deniliquin.

I spent most of my Saturday wandering around looking at cars used by Presidents from different eras. After a while I started getting a stiff neck from continually gazing up at the intricate detail of various full-sized aircraft that were hanging from the ceiling. Alongside most of the items on display was a glass faced box, framing a description of the article. For some of the larger displays, this box would have a button to press, this would activate a recording with an expanded description that I could listen to.

My attachment to the Henry Ford Museum may have come through my employment, but the experience was more than that. I saw some of my own history represented in the displays and all of them sparked a flood of memories. Walking out of the doors at closing time I had only seen part of the display and yet, I felt enriched for the experience.

Looking forward to another dose of Ford history, next morning found me embarking on a four-hour visit to "Greenfield Village." More than just a collection of buildings from the past, these were structures that had been rescued, dismantled, moved and rebuilt in the village. Along with homes from the famous, were more humble dwellings showing how early Americans lived.

Old Henry's birthplace held a special interest to me and seeing it, I could draw parallels with the Pizzey-Mud home my father had built for us. I explored the first Ford Factory and although I knew how the old manufacturing process worked, it still piqued my interest. I took quite some time resolving how everything progressed until a complete vehicle rolled out the door. His office, including the desk complete with a swivel chair and telephone, looked as if he could appear there at anytime and begin making a phone call, or dictating a letter.

In the last few minutes of my tour, I found another display dedicated to remembering the importance of another of my heroes. Thomas Edison's factory may have been small. However, the curators had determined it was central to the industrialisation of the country and it stood tall among the other exhibits.

In the afternoon I toured the Renaissance Centre, a huge building of at least five towers with one tall column in the middle. I understand that the city fathers had approached Ford to see if the company would be interested in a futuristic building complex, which they envisioned as covering a full city block. After some time, Ford agreed and the plan was undertaken.

It's a fantastic structure. Entering the foyer of the Detroit Plaza Hotel, my eyes were drawn up, taking in the volume of the space. The void rose about eight floors and I found myself looking up as I walked toward the reception area. As I explored the hotel, I found myself reaching out, touching the finishes of the fabrics of the décor in the features of many different ballrooms, restaurants and meeting rooms that made up the complex. I learned that some Ford Motor Company executives had occupied apartments on the top floor of the tallest of the buildings.

To have an opportunity to spend two days exploring my employer's history filled me with pride. I would highly recommend to anyone travelling to the United States, that they put Detroit on their itinerary and experience the same places I did. It would be one very rewarding adventure.

CHANGES

The mid 80's had witnessed great improvements in the methods employed by our service department, so much so that I had become almost redundant. My team had been carrying out their normal duties, handling difficult customers and solving complex service issues without much need for my personal involvement. Leaving me free to find ways to further streamline and improve our operation. Only when Noel or one of my seniors from overseas specifically requested my input, would my normal business of overseeing everyone be interrupted.

However, while my role within the service department was going along smoothly, I had become unhappy with my situation at home and I could see it beginning to affect my work. Something needed to change and as much as I loved my role within the company, I felt my health would suffer if I didn't make a change. These personal issues had made me distant and vague. Although I needed a change, it took a few weeks to realise what had to be done.

Discussing it with Noel, we agreed that I would move out of the National Service Manager's role and take up a sales position in Zone One. There I would be covering our tractor dealers from Rockhampton to Cairns.

Zone One proved an interesting time. Our tractors and other machinery suited the market and we were selling plenty. This type of agriculture bred a different kind of farmer, these were men and women who knew the meaning of toil. They were hard negotiators, yet people who would offer a helping hand to anyone down on their luck. I would never be one of them, and yet they made me feel they were my people.

Based in Queensland and representing Ford in Zone One proved itself a great time to be alive and because of it, I decided to make my home on the Gold Coast permanent.

During my time in Zone One, rumours of Ford's takeover of New Holland became a reality, Noel spoke with me again, asking me to consider returning to manage the Service Department in Broadmeadows. He knew the challenges the new company would face, and said he needed someone he could trust steer the team through the process. His case was enticing and he strongly pressed for me to return.

Frankly, I could not accept the idea. All of the scuttlebutt that I had been hearing from those close to New Holland Australia's management in the field, made it clear in my mind that a return to Melbourne did not appear to be in my best interests. Eventually, a small number of our tractor team relocated from Broadmeadows, to Cranbourne. On Friday the office was humming and by the following Monday morning our Sydney Road office had been hollowed out. Later I learned from several members of the tiny group of Ford Tractor Operations employees, that when they walked through the front door of New Holland in Cranbourne their excitement dissipated. They were greeted by a wall of at least one-hundred solemn faces.

Within a very short time, it became quite apparent that New Holland staff were firmly of the belief that their company had bought Ford Tractors. The way the new company operated and their different operating systems were nothing like the logical and easy to follow methodology that I was used to. Ford and New Holland had similar histories when it came to serving a farmer's needs. However, the way their Australian arms operated were completely different and if I found it confronting, so did my Ford colleagues.

Noel Howard, from his first day of running Australia's Ford Tractor Operations, clearly demonstrated his management capability. Although it could have cost him his job at one time, the development of the Ford 8401, and subsequent sales of seven hundred and thirty-nine units. It was holding market leadership through the eighties that proved it, not to his colleagues, but those in World Headquarters too.

However, his success had made enemies within the company and in the end, no matter how close he was to those at the top, his job was now in jeopardy.

New Holland had many wonderful products, but their manner of managing their new Ford New Holland business, was, in my view, totally different to the Ford I knew and loved. Those in charge never seemed to have an overall plan or an idea to create reasonable profit margins. To Ford Tractors staff, stock control or forecasting that we were used to appeared vacant. We believed these tools and the skills of our people who used them would be required to sustain the new company. Our predictions could not have been more wrong.

Passing on taking the Service Manager Tractors role, I continued working in Queensland as a Zone Manager for Ford New Holland.

AN EVEN BIGGER CHANGE

April 1986, and with my personal life in a much healthier place, I remarried. Excited to settle down with a fresh start to life, my marriage to Gwen came with an added bonus, her twelve-year-old son, Anthony.

Initially, we lived in Gwen's high-rise apartment at Seacrest, Surfers Paradise. Overlooking the sea, the view and the cosmopolitan vibe of the city was wonderful. However, with three of us occupying an apartment, it became obvious we needed something bigger. Before long we found ourselves moving into a new house at Ashmore. We remained there for a few years before buying a much larger house in Parkwood, near the Gold Coast University Hospital.

During 1992 I had become tired of my years of constant travel and wanting to enjoy spending more time with my family, I decided to resign from the tractor industry. Taking an early retirement, we were still living on the Gold Coast and the relaxed lifestyle suited me.

However, having been on the go for most of my working life, I was beginning to become restless and, there is only so much gardening a man can do. I needed an interest. As it had in the past opportunity soon came knocking again and I found myself being approached by a chap living on the Coast, asking me if I would be interested in working with him.

After some discussion about the role and even more interrogation about his company, we agreed to terms and a new adventure commenced. I would be selling Waste Water Treatment Plants around the many emerging hobby farms and other rural areas in and around the Gold Coast. These were a new type of system. Employing a different approach to treating sewerage, they are the next stage up from septic tanks. The places where we sold and installed them, were in zones outside of local government sewerage treatment areas.

Similar to every other selling or service role I had worked in, my daily plan involved travelling within a radius of one hundred kilometres of home. I was always meeting prospective customers who had made an enquiry through the company office. I continued working with them for about three years and over that time sold three hundred and ninety-six systems. It would have been nice to sell another four and even out the tally, but I knew it was time to go.

RETIREES HAVE TO DO SOMETHING

Like most grey-haired Aussies living the retirement dream, it was great to have some down time and begin adjusting to quieter week days. To take advantage of all this leisure time, we purchased a caravan and made time for touring around the local countryside and enjoying the scenic beauty of our great country.

While thinking about and planning our retirement, I received an email from Rory Day, then editor of "Classic Tractor" magazine, enquiring if I would be interested in writing an article on the development and history of the FORD 8401 tractor. Knowing his magazine for its wide circulation and respect within the industry, I asked him to call me. After asking a lot of questions about what he needed the story to say and about how long to make it, we agreed that I would write about the 8401's development and the reasons behind it.

When I read the article in the July 2012 edition, I was not only proud of my writing, but being part of Noel Howard's visionary team of engineers who out of a need to meet the market, built and developed a tractor that captured the 100-horsepower market segment. The importance of this tractor should not be underestimated as it helped Ford Tractor Operations Australia reach market leadership during the early part of the 1980s.

Looking back, the article clearly showed how Noel's plan unfolded and how the tractor developed. In association with the Basildon plant, the production of Ford 8401 tractors reached seven-hundred-and-thirty-nine units. This was a big number in a falling tractor market.

By now we had Anthony finishing school and had even more time to travel. It coincided with Anthony beginning work for a local pool shop owner. For Anthony, it meant days spent on the road, maintaining pools for their customers.

Anthony's standing with the business owner and his team at the pool shop, led to him asking me if I would like to come on board, too. Having toyed with the idea of finding something to do, his timing could not have been better and soon I was working for the same company.

My new job involved looking after their second shop close to the Central Business District of the Gold Coast. Although water had been a theme of my life from my earliest of years, this was different. Using a spade to open and close channels as a boy, I understood irrigation and had sold water treatment plants. Pursuits, which were all born of necessity.

In the pool shop, I saw a different side of selling, different to tractors, or anything I had done before. The average pool owner was a leisure industry customer and although they might have haggled a bit, most of the time the money being spent wasn't included on their balance sheet. It was a discretionary spend. I enjoyed the banter and occasional barter with our customers. To me, I found it very interesting work and besides, it kept me busy.

In 2014, Anthony had been offered an opportunity to purchase the main pool shop based in Nerang. With it came the Parkwood shop. I continued to keep busy in the Parkwood premises and stayed working with my son for a couple more years. It took a lot of negotiation and haggling with different institutions before a deal could be agreed. However, Gwen and I looked on with pride as Anthony took the keys to his very own business.

Since buying the business, Anthony's King Neptune Pool Centre has grown considerably. He has a happy group of people working with him and I am pleased that he has a number of Ford vehicles making up the main part of his fleet.

Every time I drive past, or see the large Metricon Stadium on television, my heart skips a bit. This huge organisation trust Anthony to maintain their pools. Not only do they have spas, and internal pools, they have another on the edge of the playing field. A lure for television, this pool is a venue that patrons can stand in while watching a match during hot summer days

A year before buying the King Neptune Pool business, Anthony and his partner Natalie had purchased a home in a new housing area called Bonogin. Being situated twenty-five kilometres south of Surfers Paradise, and only ten minutes from the large Robina Shopping Centre, it has a superb outlook and facilities to match. Built on a picturesque one and a half acres, the family had plenty of room to grow.

This prompted Gwen and I to talk about selling our Parkwood home and build a comfy Granny Flat at the rear of Anthony's home. The young couple had suggested it many times before and now Gwen and I were ready to move. It's a great initiative as we have the security of being close to family. For Natalie and Anthony there were benefits too, they could rely on us to be around for any grandparent duties that arose.

Being away from a main city area for the first time since leaving Wakool, I found it amazing, the area is very quiet, no major traffic noise, very scenic and only the warbling of the kookaburras in the morning.

Anthony and Natalie have given us a wonderful grandson, Kai, who spends a lot of his free time with Gwen and I. At just seventeen he has a much wider view of the world than I did at his age and will chat with us on a great range of topics. He continues to amazes us with stories about his interests and escapades, things that are way and beyond what I was doing at his age. I guess it's probably called progress.

The move has exceeded our expectations and contrary to early doubts we might have had, we all get along fine in our bush surroundings. It could be that the great peaceful area makes it easy for us to live together. Because we are up in a hilly area, there is no flooding after heavy rain and even on the hottest of days an easy living breeze sweeps across the dam and up the valley. Our very own little patch of paradise is only fifteen minutes' drive to the largest shopping centre on the Gold Coast.

Although many years have passed since I worked for Ford, I still manage to communicate with a number of workmates and colleagues from Australia and overseas.

We were lucky that in 2016, Mike Hutton from the United Kingdom came to visit. Mike stayed at home with us for a week and it was as if time had stood still. We told stories, swapped lies and generally had a great time. Being with us for a week meant there was plenty of time to show him around and plenty of time for recalling our old victories. After leaving us, Mike went onto Brisbane where he had planned to catch up with about forty relatives that he had found on the internet before leaving the UK.

If I had my time over again, it would still be another wonderful experience to join the Ford Motor Company and do it all one more time. Besides the great opportunity I had in seeing Australia and many other parts of the World, it also was wonderful to have been in a position whereby, over time, I had eighty-three cars assigned to my use. During my twenty-nine years with Ford, I seldom had a vehicle that reached 30,000 kilometres prior to them being "Grounded." I didn't keep a full record of the distances I travelled, but between October 1965 and June 1966, I remember totalling 62,000 air miles and 30,000 road miles. As a Zone Manager my mileage would have been around 3,500 miles per month and that lasted for about seven years, which meant I managed to drive just under 300,000 road miles during that period.

In my time with Ford Tractors, I travelled overseas on eighteen different occasions, and to think that I have been involved in many great projects from the very beginning and seeing them flourish in the market place, fills me with enormous pride. Not only did my move from Massey Ferguson to Ford offer me an opportunity to rise through the ranks of management, I also took much pride in being able to assist our dealers to put better systems in place. We were geared to train dealers and their staff in better managing their service departments and seeing them as profit centres.

Not everyone has the privilege of seeing great places around Australia and to travel the world as I did in my role with Ford. Although the travel was always for work, I believe there are not many companies who treated their executives the way our company did and for that, I feel extremely privileged.

WINNERS ARE GRINNERS

It's often argued that when we retire many people wonder what they will do with all the spare time. I found retirement to be far from what most of us expect. My first reaction was, when there is so much to do at home, how did I ever find time to go to work five days per week?

Naturally, there is supposed to be time to fix things around the home that had long been put off, especially when one has been as deeply involved in a normal job as I believed I was. On the bright side, now there was time to look at doing things that are possible with all the extra time available.

Our priorities had changed, there was time to travel around Queensland in our caravan. With a new set of priorities, we found new opportunities to stop and explore the wonderful attractions of this great country of ours. For me, it was as if the places I had been or even just whistled through before were revealing themselves for the first time. The eyes of a tourist seemed to be tuned differently to those of the commercial traveller.

One attraction on the northern end of the Gold Coast is the number of houseboats that are available for hire. Most will carry up to twelve people and on these vessels, you can laze away your time sightseeing around the many canals. There is time to relax, do some fishing, and talk to your friends. Gwen and I have enjoyed many of these jaunts over the past twenty years.

Another pastime I have found interesting, is the intriguing number of free competitions that are available on the internet. Some of these offer various prizes even some offer items of considerable value. I understand some can be scams and do my best to avoid those, and as many people enter these competitions I do not expect to win often.

Although it's only time I'm investing, as the saying goes, if you don't try you can't win and I have been fortunate on a few occasions by winning a handful of minor prizes.

One competition that stands out in my memory is a trip for two to Perth, by JetStar. Winning it was quite a thrill. So, with our bags loaded we headed to the airport. On our arrival in Western Australia, Gwen and I decided to hire a motorhome and take our time driving down to the historic town of Albany, stopping at Cape

Leeuwin to see the lighthouse. It's not far from Augusta and to watch the power of nature where the Great Southern and Indian oceans meet is a moment I'll never forget.

We had planned on travelling around the coast until arriving back in Perth. Margaret River was beautiful and certainly gave us reason to stop overnight. Whilst I had seen much of the area during my days with Ford, and because she had never visited Western Australia before, it was a new experience for Gwen.

Another free to enter competition we won, had been run by Queensland Tourism and was a trip to Longreach. We travelled overnight by train being accommodated in a sleeper cabin before arriving at our destination. The trip took twenty-four leisurely hours and we made good use of the bar before taking our place in the dining car. There we took our time selecting our restaurant quality meals from their menu.

At Longreach we checked into our luxury accommodation where we stayed for five wonderful days. Before setting out to explore the area, we had agreed on visiting the Qantas Museum, The Stockman's Hall of Fame, and elected to book an evening trip down the Thompson River. It was a top week away and one where we thought Queensland Rail had excelled themselves. Although we had accumulated a raft of yarns to tell our friends, we were tired and happy to finally be travelling home by plane.

Another day I had been visiting a local bottle shop intent on picking up some liquid vitamins. At the checkout, the lady working the till recognised me. Saying she had seen me in there a couple of times before, she gave me four tickets and asked if I would like to enter a competition. In an offhanded way she told me to just fill them out and put them in the post, something I did on the way home.

A few days later I received a phone call from Sydney with the news that I had won a trip for two to visit Tahiti for seven days.

Gwen and I flew to Sydney and boarded our plane to Tahiti. We found ourselves sitting beside a friendly couple, who were excited to be on holiday too. This was another opportunity to meet new people and we all engaged in a happy chat that lasted until we touched down.

Our vacation in tropical French Polynesia felt as if it flowed in an endless and easy seven-day rhythm that finished too soon. Our hosts proved their image an easy going and warm people is correct. Even now, our memories of this paradise drift into our conversation often.

When I think about winning, I am reminded of the 1982 AMERICA'S CUP campaign and how we shared a lot of good-humoured banter with our contemporaries in Michigan. When the salt-spray had settled, one of our Ford Tractors Australia team came up with the following cartoon to finally rub some of that Newport salt into their American pride.

These are just a few of many mementos and photos saved from my trips abroad and at home. These all have a story and I keep them in 'Fred's Corner,' an area set aside in our living space that's dedicated to my achievements.

Fred, happens to be an affectionate nickname my family has bestowed on me. I've grown used to it now.

ACKNOWLEDGEMENTS

None of this story would be possible if it were not for Milton Bayfield, my science teacher. More than just an educator, Milton was a man who taught me how to think by always posing intriguing questions like:

> *"When you place a kettle, filled with water, on a hot stove, why does it boil after a few minutes?* or, *"The ship, Queen Mary, weighs over 80,000 tons, and is basically made of steel, so why does it float in the sea?"*

My agriculture teacher, Ian Hore became another important influence in my life. During my high school years, Ian, led me through the basics of farming and suggested I join a Junior Farmers Club to further my knowledge of farming practices. I heeded his suggestion and it proved a good grounding for my future and, as he predicted, became an excellent move.

Jock Berry, was the Chief Engineer at Massey Ferguson Test Division when I first started there and it was his valuing me as an equal that had the biggest influence on me. By exchanging information between his understanding of engineering practices and my knowledge of farming, we both benefitted. However, I think I received more information from him, even if he was often surprised with my stories of farming and their ingenuity for finding practical solutions to breakages.

Mike Hutton, my friend and Senior Ford service trainer at Boreham House in the United Kingdom is a very dedicated tractor man. He could always cover any question relating to the company's tractors. Mike proved himself an inspiration to many of Ford Australia's tractor dealers when he came out to conduct service training in a number of different locations.

Noel Woodford, our in-house legal counsel while I worked for Ford. For over twelve months, Noel gave me basic lessons on methods I could find useful to cover possible legal questions from tractor owners. Difficult questions that could arise at any time. His time and patience gave me an ability to remove emotion from a problem, a virtue that has helped with life outside the tractor business, as well as inside it.

It is impossible to list the many in house Ford Australia training staff, who continually conducted sessions on how to improve the number of day-to-day challenges that arose in our dealings with customers. Each one sharing their knowledge and experiences generously, so we all benefitted. These fellow workers made me proud to be part of the team.

Noel Howard the last General Manager of Ford Tractor Operations Australia. This man was an incredible operator, who encouraged and led his staff by example. A friend and colleague who was an inspiration not only to company staff, but also to our dealers. Noel was a real innovator, opening up new ideas, and using his capacity to open doors in Federal parliament for the betterment of the agricultural industry.

It was Noel who nominated me for the Australian Service Manager position, and over the decades, we spent many hours together, always looking for ways to improve the company's goals. Our discussions were sometimes robust with very strong opinions put, but the outcome was always in Ford's and our dealers' best interests. It was more than a pleasure to work with him, as well as being a close friend to him and his family, for over fifty-nine years. I will never forget him.

David Hosking, you are one of my most inspiring workmates. A colleague who came up with new ideas that always worked. I have always been amazed at your ability to understand my verbal concepts and then put them into print that made sense. Yours is an unmatched talent.

When Ford staff transferred to New Holland at Cranbourne, you declined to go and transferred to Ford's national parts depot in Broadmeadows. However, your talents were soon noted by head office at Ford Australia. Promoted to the light truck division as manager, within a short time your efforts resulted in major increase in the sale of ford utilities. It would be remiss of me to forget to mention the input of ideas you contributed to the design team when they were planning the very successful Ford Ranger series.

My co-author Terry Probert, for the incredible time he has spent on this project, the suggestions and all of the phone calls taken to get this publication into a readable form. Terry was never short on ideas and the humour that arose on many occasions kept me going till the end.

Brandon Brooks, a tractor collector and farmer from Hamely Bridge in South Australia. Many thanks for the photos used in the cover. Showing an 8401 Ford tractor working in the field and for supplying all of the other photos for us to choose from.

Ian Denton, my friend. Ian is a man who has spent most of his working life as a Ford Australia tractor dealer, servicing the needs of farmers in and around Farrell Flat, in South Australia. Thank you for all your service assistance over the years and for providing out of print information on old models to me and our customers.

Your ability to find ancient service bulletins and service information hidden in manuals that your family have stored for years, became invaluable. This information was often unavailable anywhere else in my Ford world. Yet because you and your staff had been strong Ford dealers and supporters for most of the time Ford had been selling tractors in Australia, you were able to share items from your archive.

Ian, I will carry those memories and the value of your friendship with me all of my days.

The many dedicated Ford tractor dealers that I met and worked with over the years. You are too many to mention individually, but by supporting the company by servicing the needs of farmers and industrial businesses, you helped make my job a pleasure. A big thank you must also go to the thousands of customers that were so supportive to Ford Tractors in the good times, times that have now long gone by.

Looking back over my life, I hope that I have at some time, somewhere, and somehow, contributed something beneficial to someone in this crazy world of ours.

Along the way I have picked up a couple of sayings I have found helpful during my career and these are:

- The solution to most situations is common sense.
- Stay away from negative people, they will always have a problem for every solution.
- Never criticise anything unless you can offer a much better solution.

AUTHOR'S NOTE

For me, writing my memoir has been an enjoyable task that has filled my days with wonderful reminiscences of a vibrant and developing agricultural industry. Allowing me to sleep the dreams of a happy man.

I sincerely hope that readers of these words find as much enjoyment in the stories from my memory as much as I have while recording them.

Ian

CO-AUTHORS NOTE

1969 and with the April sun setting, my balmy Hobart evening was being brightened with orange and red flashes tipping the lolling waves on the Derwent River. Checking into my hotel, the receptionist told me another Ford Motor Company employee had asked after me. She smiled and suggested I should meet him in the dining room bar once I had settled in.

Easy to recognise in my Tractor Demonstrator uniform, I was approached by a darkhaired man who could have walked straight out of a Hollywood film set. Ian O'Rourke made that kind of impression on people. Over dinner we began a long and lasting friendship, one which has culminated in this story about his life.

Although life as a tractor dealer in the eighties didn't bring the rewards I had wished for. The example shown by Ian and his colleagues has led me to become a story teller. By helping Ian with his memoirs, I have become better at my craft. It is not easy to do, but I hope that Ian's stories encourage more tractor people to write about their history too.

Ian, thank you for treating a nineteen-year-old, wet behind the ears, Ford Tractor Demonstrator to dinner on that balmy Hobart night, and for trusting me to help with your project.

My hope is that my literary effort has helped to repay that ancient favour.

ABOUT TERRY PROBERT

Terry Probert is a novelist and biographer who lives with his wife Ruth in Bendigo. Following a career in the Auto and Machinery Industry, Terry chose to put his love of story into book form. First with KUNDELA an adventure novel based around an older hero and set in South Australia it received a Commended in the 2013 Christina Stead Awards.

Other publishing credits include a murder mystery, Voss, The Price of innocence. This was followed in 2019 with Gillespies Gold, another adventure that followed Kundela's characters trying to find a cache of gold lost in rumour and mystery.

The Covid-19 years have produced two biographies, Ian O'Rourke's Memories of a Tractor Man and, NOEL the biography of Noel Howard the last general manager of Ford Tractor Operations Australia.

For 2022, Terry has two Voss novels to complete along with Toby Farrier, a modern teenage adventure to get ready for publishing.

Made in the USA
Monee, IL
13 July 2026

57049341R00174